EARTH IN CATACLYSM

By Philip G. Budd

Copyright 2014

ALL RIGHTS RESERVED

No portion of this book may be used in any form without written permission of the author, with the exception of brief excerpts in magazine articles, reviews, etc.

Title ID: 4508101

ISBN-13: 978-1493654291

Cover Photos by Janie Johlman

Castle Geyser (front) & Lower Falls (back)

Yellowstone National Park

Book Title by Anne Habermehl

Acknowledgements

I am grateful to Ed Boudreaux, Robert Lawrence, Rob Bracken, Jay Wile, David Bergman, Bob Compton, Brent Carter, Harald Heinze, Ed Holroyd, Frank Gauna, Michael Oard, and Larry Hamilton for donating their time and expertise by conducting peer review and constructive criticism of various manuscripts and sections of *Earth in Cataclysm*. My thanks to the participants of the In Jesus' Name Flood Model contest and its moderator Joe Bardwell for their insights and critiques. I am also grateful for Kyle Budd's computer support efforts. The helpfulness and self-sacrifice of the individuals above does not necessarily signify agreement with *Collapse Tectonics Model* conclusions.

OVERVIEW

Theories and models are useful scientific interpretation tools that can never be proven. They resemble a "Lady-in-Waiting" in a King's palace who will never become Queen. It is only a matter of time before she will be replaced by a more useful interpretation tool. *Earth in Cataclysm* was designed for that purpose. **The skeptical reader will discover the Collapse Tectonics Model is favored over its Plate Tectonics predecessor exclusively on the basis of scientific accountability.**

Objective analysis requires setting aside preconceptions to evaluate the evidence without prejudice. There is no need to force science to accommodate Scripture or to force Scripture to accommodate science. *Earth in Cataclysm* examines how all direct Scriptural information can be effectively incorporated into the construction of a geomorphic model. Geologic evidences alone independently affirm the Scriptural Account without first accepting Scripture as truth.

Geologic evidences reveal much about chronological sequencing - but little about geologic time. An Earth Age of several thousand years is revealed to be potentially compatible with Rock Record evidences. For archaeological correlation purposes, speculative dates from the oldest Septuaguint (LXX - Old Testament Greek version) geneaological record are assigned. However, that time frame is subject to an accordion-like expansion or contraction of interpreted dates at the reader's discretion.

A fresh portrait of Earth's cataclysmic past emerges when individual topics are evaluated collectively. Geologic features and characteristics from around the world provide comprehensive model context supporting the Collapse Tectonics Model. Contrary assumptions essential to Plate Tectonics, Radiometric Decay Dating, and the Geologic Column are effectively challenged in the Addendums that follow. The text of each Table of Contents topic features an evidence, description, or argument characterized by a **bold sentence.**

TABLE OF CONTENTS

WORLDWIDE FLOOD MODELING   1
GOD'S EYEWITNESS ACCOUNT   2
    Creation Week Implications   2
    Antediluvian Climate   3
    Antediluvian Crust   4
    Antediluvian Geothermal System   6
    Worldwide Flood Account   7
    Biblical Genealogies   10
    The Mystery of Mankind's Ancient Past   11
PRECAMBRIAN VERSES PHANEROZOIC FLOOD   12
    Worldwide Flood Deposit Thickness   13
    Time of Deposition   14
    Unconformity Context   15
    Metal Ore Lithology Changes   16
    Precambrian Stratigraphy   16
    Post-Flood Erosion and Deposition   18
    Sinking Continental Basins   19
    Body Fossil Transport   21
    Post-Flood Recolonization   22
    Fossil Reefs   23
    Halite Formations   24
    Air-breathing Land Animals   27
    Post-Flood Calamities   28
COLLAPSE TECTONICS MODEL   30
COLLAPSE TECTONICS MODEL CHRONOLOGY   31
    Magma Types   32
COLLAPSE TECTONICS SUMMARY   33
MAGMA SEGREGATION AND MIGRATION   38
    Pyroclastic Volcanics   41
    Pyromagma Vug Formation   42
    Crystal Melt Mush   44
    Antediluvian Crust Solidification   45
    BUDD'S PRIMORDIAL EARTH MODEL (by Boudreaux)   45
    Pre-Continental/Pre-Oceanic Crust Comparison   49
    Pre-Oceanic Crust Destabilization   51
    Triggering the Worldwide Flood   51
    Deep Ocean Trenches   52
    Proterozoic Sedimentary Deposition   54
    Porous Crust Collapse Potential   55

OCEAN BASIN COLLAPSE   59
       Orthogonal Seafloor Topography   61
       Paleomagnetic Anomalies   62
       Island Arcs   63
       Pockmarks   66
       Seafloor Basalt Intrusion   68
       Submarine Canyons   69
       Raising Continental Rims   71
       Coastal Monocline Uplift   73
       Mid-Atlantic Ridge   74
       Archean Cratons   77
       Worldwide Flood Heat Dissipation   78

POST-FLOOD CONTINENTAL CRUST COLLAPSE   80
       Earthquakes   82
       Sumatra-Andaman Earthquake   85
       New Madrid Earthquakes   86
       Mud Volcanoes   89

COLLAPSED POROUS CRUST EVIDENCE   93
       Precambrian Phosphorites   94
       Precambrian Pseudomicrofossils   95
       Precambrian Stromatolites   96
       Precambrian Metamorphic Mineral Fabric   97
       Metasomatic Alteration   99
       Parentless Polonium Radiohalos   100
       Sonofusion   104
       Continental Shields   105
       Precambrian Banded Iron Formations   106

ROCK RECORD CONTEXT   107
EARTH'S CHANGING STRATIGRAPHY   109
PROTEROZOIC WORLDWIDE FLOOD   110
POST-FLOOD MILLENNIUM   111
       Stromatolite Recolonization   112
       Ediacaran Recolonization   113
       Cambrian Recolonization   114
       Paleozoic Carbonate Deposition   114
       Entombed Fish   118
       Fossil Reefs   119
       Ecosystem Reversals   120
       Hexapoda Gap   120
       Evaporite Precipitation   121
       Post-Flood Terrestrial Climates   122

  Mesozoic Ecosystems   123
  Mesozoic Clastic Deposition   124
  Tsunami Wave Backwash Deposition   126
  Cyclothem Deposition   129
  Fossils in Tsunami Backwash Sediments   130
  Bones in Fissures   131
  Erratic Boulders   133
  Cenozoic Ecosystems   134
  Mountain Upheaval   135
  Paleogeosynclines   138
  Mushroom Tectonics   142
  Metamorphism of Mountains   144
  Continental Basalt Flows   145
  Continental Rift Systems   146
  Clastic Dikes   149
  Batholiths   150
  Calderas   153
  Limestone Caves   154
  Carolina Bays   155
  Ecosystem Migration   157
  Ice Age Climate Change   158
  Ice Age Extinctions   161
  Outburst Floods   163
  Dinosaurs in Permafrost   165
  Antarctic Ice Cap Formation   166
  Ice Age Ends   166

POST-FLOOD DILUVIAL FEATURES   167
  Water Gaps   168
  Planation Surfaces   168
  Sheetflow Conglomerates   169
  Pediments   170
  Grand Canyon Stratigraphy   171

ECONOMIC APPLICATIONS   175
  Ore Mineral Intrusion   175
  Placer Deposition   176
  Coal   177
  Petroleum   178
  Carbonate Exploration   180
  Clastic Exploration   183

COLLAPSE TECTONICS CRITIQUE   185

ADDENDUM 1: PLATE TECTONICS   186
    Mantle Convection Cells   187
    Earth Age   188

ADDENDUM 2: RADIOMETIRC AGE DATING   188
  CARBON-14 DATING   188
    Earth's Decreasing Magnetic Field   190
  LONG HALF-LIFE ISOTOPE DATING   191
    Neutrino-induced Fission   193
    Plasma State Accelerated Decay   194
    Piezonuclear Fission   196
    Critical Mass   196
    Recent Lava Flows Yielding Old Earth Ages   197
    Parent/Daughter Ratio Alteration   198
    Isochrons   200
    Naturalist Earth Age Estimates   201

ADDENDUM 3: THE GEOLOGIC COLUMN   202
    Worldwide Unconformities   204
    Paraconformities   205
    Uniformitarianism   205
    Precambrian/Tertiary Interfingering   206
  BIO-STRATIGRAPHY   207
    Transitional Fossils   207
    Cambrian Explosion   208
    Iterative Evolution   208
    Overthrust Belts   209
    Reworked Index Fossils   210
    Dragons Amoung Us   211
    Man Before His Evolutionary Time   215

REFERENCES   216

DEFINITIONS   232

WORLDWIDE FLOOD MODELING

The challenge for Flood Worldwide models resides in identifying reasonable cause-and-effect mechanisms to harmonize Scriptural information with geologic evidences. Failing Flood Models propose postulates and parameters that are either physically inadequate or that require miracles not specified by Scripture to accomplish the attributed effect. *"Miracles are historical events, but it is inappropriate to invoke miracles just to make sense of the data"* (Tyler, 2006, p. 75). The Worldwide Flood Account only specifies miraculous intervention by God bringing creatures to enter the Ark (Genesis 7:15), shutting the Ark door (Genesis 7:16b), and possibly sending a wind over the earth (Genesis 8:1b). None of those miracles would have affected Rock Record evidences. Miracles not specifically referenced by Scripture are possible. However, miracles circumventing the laws of physics automatically forfeit scientific accountability.

Noah's Ark was God's earthly provision to survive the Worldwide Flood. Scripture provides no reference indicating God's miraculous intervention to preserve the Ark voyagers from life-threatening side effects other than the floodwaters. For example, an author's declaration of a heat-dissipation miracle to avoid boiling the oceans dry and sterilizing the planet would not be supported by science or Scripture. Science can not independently affirm the Scriptural Account if an author declares a miracle for the purpose of rescuing an otherwise unreasonable model parameter or to nullify an adverse side effect.

Pertinent questions must be asked and answered to derive reasonable conclusions.. How did Earth's primordial climate and crust differ from today? What was the source of the floodwater? What process might initiate a Worldwide Flood? What caused the floodwater to suddenly inundate Earth's surface and then rapidly regress off of the land? What segment of the Rock Record most reasonably corresponds to Worldwide Flood deposits? Do Earth's landforms and stratigraphy provide comprehensive, properly sequenced evidence supporting model conclusions? Can the Scriptural Account be applied as legitimate historical information?

GOD'S EYEWITNESS ACCOUNT

When ancient accounts lead to a superior scientific explanation of the geologic evidence, the reliability of that information is affirmed. However, Scripture is not allowed to justify itself. Instead, the geologic evidence must independently demonstrate the scientific veracity of the Scriptural Account. Biblical information regarding Earth Age and the Worldwide Flood is very limited. So let us examine relevant Scriptural passages and select extra-Biblical accounts to avoid premature assumptions regarding what Scripture may or may not intend.

Creation Week Implications

Was there a built-in potential in creation for the Worldwide Flood? God created man with the potential to sin and fall from grace. Was the Creation correspondingly designed with a potential to fail? Or did God miraculously intervene by altering physical processes to cause the Worldwide Flood?

God pronounced the creation *"very good"* on Day 6 of the Creation Week (Genesis 1:31). The Hebrew "Towb" (2895) translated as "good" encompasses a variety of beneficial effects stemming from *moral goodness*. God's handiwork of Creation Week Days 1, 3, 4, and 5 were each described as *"good"*. That establishes context relating to each of those Days. However, the Earth-forming and atmosphere-forming activities described on Day 2 are not described as *"good"*. Perhaps God foreknew the wickedness of man would increase until His judgement would be necessary. *Moral goodness* must also allow for righteous judgement. The omission of *"good"* describing Day 2 is potentially explained as "setting-the-stage" for the Worldwide Flood at a later time. To preemptively assume the potential for a Worldwide Flood was not incorporated into the Creation Week would be inappropriate.

Did the curse associated with Adam and Eve's fall from grace (Genesis 3) initiate physical processes leading to the Flood? Let us hypothetically assume *"very good"* in Genesis 1:31 refers to a perpetual primordial state. In that case, God's miraculous intervention would be required to alter that state. However, God did miraculously intervene by imposing a curse.

"Cursed is the ground because of you ...
By the sweat of your brow you will eat your food until you return to the ground
since from it you were taken; for dust you are and to dust you will return"
(Genesis 3:17-19, NIV).

The immortality of Adam and Eve may have been directly linked to the *"ground"*. That allows the possibility that the primordial *"very good"* state was subsequently altered by the curse. Geomorphic processes may have been initiated by the curse that progressed during the Antediluvian Age and subsequently culminated during the Worldwide Flood.

Scripture permits the possibility of either a Creation Week *structural mousetrap* or a *curse-induced geologic process* that initiated and accomplished the Worldwide Flood. **Therefore, God's miraculous intervention via exceptions to the laws of physics is permitted but not essential to Worldwide Flood events.**

Antediluvian Climate

"For the Lord God had not sent rain on the earth ...
but streams came up from the earth and watered the whole surface of the ground"
(Genesis 2:5b, 6 NIV)

What conditions would allow the Antediluvian World to function without rain? Modern weather patterns are powered by the large-scale differential heating of continents and oceans that generate strong winds. Modern albedos (solar reflectivity) varies from 0.1 for forests, 0.2 for oceans, and 0.45 for sandy deserts (Byers, 1974, p. 33). However, an Antediluvian land-and-lakes topography would have prevented large-scale differential surface albedos. An absence of continents and oceans would have generated atmospheric temperature inversions, suppressing wind, rain, and rapid temperature fluctuations.

After the Flood, God told Noah, *"I have set my rainbow in the clouds, and it will be the sign of the covenant between me and the earth"* (Gen. 9:13, NIV). Rainbows form by the refraction and reflection of the sun's rays in condensed atmospheric water droplets.

Atmospheric condensation is produced by rapid atmospheric cooling. Noah may not have seen a rainbow prior to the Worldwide Flood. Temperature variations may have been too subtle to "over-saturate" the Antediluvian atmosphere and condense water droplets, thereby potentially preventing rainbows and rain.

Atmospheric dust particles provide condensation surfaces for raindrops to form. However, gentle breezes coupled with rare or nonexistant volcanic extrusions and prolific surface vegetation may have prevented significant airborn dust in the Antediluvian atmosphere. Negligible volcanic activity implies the absence of deep faulting, which conforms to the concept of an impermeable Antediluvian crust. **As an interpretational option, widespread habitable subaerial ecosystem conditions without rain would seem reasonably possible in conjunction with worldwide land-and-lakes topography.**

Antediluvian Crust

"On the third day you ordered the waters to collect in a seventh part of the earth;
the other six parts you made into dry land"
(Ezra, II Esdras 6:42a; a.k.a. 4 Ezra)

Second Esdras was retreived from *"the Old Latin Translations of the Old Testament, made from the Septuagint"* (Lumpkin, Joseph B., 2010, p. 567). This apocryphal writing is not regarded to have the same authority as the modern canon of Scripture. However, its author is credited with having written the Old Testament Book of Ezra. Because God spoke to his prophet Ezra, the verse above should be regarded as a potential elaboration regarding God's truth. This extra-biblical verse implies Earth's primordial surface was one-seventh water and six-sevenths dry land. Ezra's elaboration implies there were no continents or oceans. Instead, the Antediluvian landscape would probably have consisted of gently rolling hills and shallow lakes.

"I (God) *establish my covenant with you: ...*
never again will there be a flood to destroy the earth"
(Genesis 9:11, NIV).

The Worldwide Flood is designated as a one-time only event. What unique, unrepeatable geologic conditions could possibly induce a Worldwide Flood inundation and regression within one year?

"He (the Lord) gathers the waters of the sea into jars;
he puts the deep into storehouses."
(Psalms 33:7 NIV)

"Now as for the earth, the lower part of it is like unto a sponge."
(Book of the Cave Treasures, 300-599 A.D.; Sir E. A. Wallis Budge, editor)

If surface waters were separated from subsurface waters by a *"dividing barrier"*, the tectonic breaching of that barrier during the Worldwide Flood would be irreversible. A second Worldwide Flood would not be possible without miraculous intervention, which God decreeded will never happen.

"On the second day you created the angel of the firmament,
and commanded him to make a dividing barrier between the waters,
one part withdrawing upwards and the other remaining below."
(Ezra, II Esdras 6:41)

That extra-biblical text implies the primordial crust acted as an impermeable barrier between surface and subsurface waters. That barrier would have prevented the upward escape of over-pressured subterranean waters A breaching of the Antediluvian crust would potentially initiate the Worldwide Flood, causing *"all the springs of the great deep burst forth"* (Genesis 7:11b). **The sudden release of over-pressured subterranean waters provides a potential source for the one-time flooding of Earth's surface.**

Antediluvian Geothermal System

Crust overlying areas of highest heat flow solidified more slowly. That produced a locally thinner crust, which served as a subsurface collection area for low-density minerals. Halite, being a low-density, low-viscosity crystal melt mush, ascended rapidly to the base of the primordial crust, forming lenses beneath thin crust. Isostatic uplift, associated with underlying low-density halite lenses, raised low-relief hills and exposed dry ground on Creation Day 3.

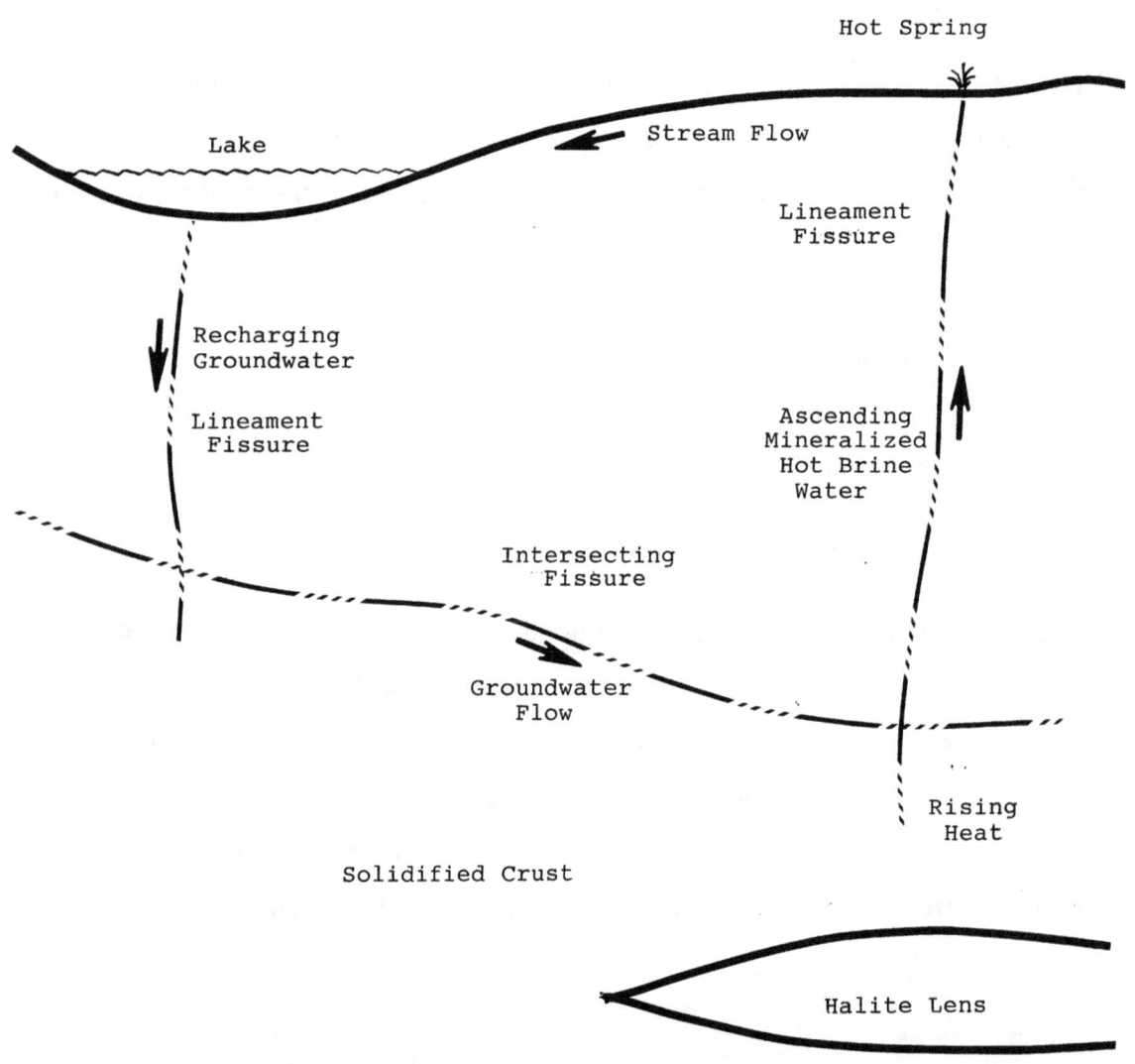

ANTEDILUVIAN GEOTHERMAL SYSTEM

The outermost Antediluvian crust rapidly cooled and shrank. Vertical cracks split apart, deepened, and extended laterally to establish an irregular surface grid of intersecting vertical lineament fissures. Continued isostatic warping of the overlying crust generated the greatest tensional flexure and deepened lineaments across hilltops.

A geothermal convection flow system, based on heating of groundwater at depth and cooling of surface water, was quickly established. Hot groundwater ascended up the hottest, deepest, lineaments, and was expelled above as hot springs on low-relief hills. Those hot springs sourced streams that flowed downhill to nearby lakes. Streams recharged adjacent soils, watering the vegetation above. Bedrock fractures and lineaments under lakes acted as groundwater recharge conduits. Cooler surface water flowed through the fracture network to deeper lineaments under hills where water was reheated to rise and be expelled again as hot springs. **A circulating geothermal irrigation system was potentially sustained throughout the Antediluvian Age.**

Geothermal systems contain mineralized waters that precipitat solids out of solution, during their ascent to the surface. Precipitation of solids is caused by lower temperature, lower pressure, and changes in acidity and oxidation-reduction potential. Geothermal systems have a natural tendency to become self-sealing, due to near-surface plugging by precipitated silica and calcite (Facca and Tonani, 1964). Excavations through soil adjacent to hot springs sometimes break precipitated mineral pipes that contain flowing hot spring water. Progressive plugging of the Antediluvian geothermal network was presumably offset by continued cooling and shrinkage of Earth's crust. Resulting tension potentially widened, deepened, and split apart fractures at a rate sufficient to maintain the circulating groundwater/surface flow system.

Worldwide Flood Account

The following text is lifted from the extra-Biblical *Frist Book of Enoch,* Chapter 65. Verses 1, 3b, 4a: *"And in those days Noah saw the earth that it had sunk down and its destruction was near ... Tell me what it is that is falling out on the earth that the earth is in such veil plight and shaken, ... And there was a great disturbance on the earth ..."* (Lumpkin, Joseph B., 2010, p. 86). This implies Earth's crust was experiencing earthquakes, and beginning to collapse and tilt immediately preceeding the Worldwide Flood. *"Even in its complete form, the Book of Enoch is not one manuscript. It is a composite of several manuscripts written by several authors. Enoch and Noah each have pieces of the book ascriubed to them ... It was not only Jesus who quoted phrases or ideas from Enoch, there are over one hundred comments in the New Testament which find precedence in the Book of Enoch"* (Lumpkin, 2010, pp. 58, 56).

".. I (God) will send rain on the earth for forty days and forty nights ... In the six hundredth year of Noah's life, on the seventeenth day of the second month (day 1) – *on that day all the springs of the great deep burst forth, and the floodgates of the heavens were opened. And rain fell on the earth forty days and forty nights ... For forty days, the flood kept coming on the earth ... The waters rose and covered the mountains to a depth of more than twenty feet ... Everything on dry land that had the breath of life in its nostrils died ... Only Noah was left, and those with him in the ark. The waters flooded the earth for a hundred and fifty days ... Now the springs of the deep and the floodgates of the heavens had been closed, and the rain had stopped falling from the sky. The water receded steadily from the earth. At the end of the hundred and fifty days the water had gone down, and on the seventeenth day of the seventh month* (day 150) *the ark came to rest on the mountains of Ararat ... and on the first day of the tenth month* (day 224) *the tops of the mountains became visible ... By the first day of the first month* (day 313) *of Noah's six hundred and first year, the water had dried up from the earth ... By the twenty-seventh day of the second month* (day 370) *the earth was completely dry"* (Genesis 7:4-24; 8:2-14, NIV). (Inserted day designations assume thirty-day months.) The context of the verses above supports the concept that all air-breathing terrestrial life outside of the ark was killed within the initial 40 day inundation period.

The Worldwide Flood lasted approximately ten and one-half months and required nearly two additional months for the land surface to dry out enough to make the land habitable. Therefore, Worldwide Flood deposition is limited to a one-year time frame.

"Where did the water come from that supplied a forty-day rainfall? It is easily demonstrated that the maximum amount of water that can be maintained in clouds in our entire present atmosphere is only about 10.54 cm for a saturated atmosphere with a sea level temperature of 28 degrees Celsius" (Byers, 1974, p. 114). *"That amounts to 4.14 inches or a forty-day rainfall rate of 0.00431 inches per hour – hardly a torrential downpour"* (Dillow,

1982, p. 63). Atmospheric water would presumably source only a tiny fraction of the floodwaters required to inundate Earth's surface. Therefore, the primary floodwater source was probably the "fountains of the great deep".

The bursting forth of the "fountains of the great deep" conforms to a sudden release of an over-pressured subsurface reservoir after breaching an impermeable crust barrier. Pressures would have rapidly decreased along a depletion curve as the contents of over-pressured reservoirs were expelled and overpressured subterranean reservoir pressures normalized. It is reasonable to assume that the bulk of contributions from the "fountains of the great deep" occurred during the initial few days of the Worldwide Flood and tapered off thereafter.

When did the over-pressured reservoir depletion curve achieve neutrality? Scripture states above, *"For forty days, the flood kept coming on the earth."* The "fountains of the great deep" potentially stopped flowing as early as day 40. The Worldwide Flood presumably achieved a minimum depth of 20 feet no later than day 40 and that minimum depth was sustained until day 150 when *"the ark came to rest on the mountains of Ararat."*

Erosional flattening of Earth's topographic highs and deposition in topographic lows would have contributed to increasing minimum water depth. However, the most impactive erosional flattening would be have occured as torrential rains flowed rapidly across land surfaces during the initial few days of the Worldwide Flood.

Fountain contributions and erosional flattening would both be highly concentrated at the onset of the Worldwide Flood. Scripture also appears to affirm the total inundation of the Antediluvian surface sometime between day 7 and day 40 of the Worldwide Flood. Consequently, it is reasonable to assume that subaerial evidences of air-breathing animal tracks and nests would be limited to the initial 40 days of the Worldwide Flood.

Air-breathing animal track and nest evidences would not have resumed until the Ark voyagers spread, multiplied, and reoccupied the land following day 370. This poses a serious logistical problem if these track and nest evidences are located on top of hundreds to thousands of meters of other sediments also being attributed to Worldwide Flood deposition. Mudcracks would not be possible during the initial 40 days of torrential rains or until day 224 when post-Flood subaerial exposure began again. Raindrop impression preservation would also have been prevented by worldwide inundation from no later than day 40 until day 224.

Where did the floodwater go when it receded off the continents? Earth's modern atmosphere contains only enough water to blanket Earth's surface with one to two inches of water. Therefore, evaporation can not account for removing water from the continents after the Worldwide Flood. Flood waters drained off the continents by raising land, deepening ocean basins, and/or by possible supernatural intervention via God's wind (Genesis 8:1).

If ocean basins existed during the initial 40 days of the Worldwide Flood, those ocean basins would be expected to exhibit the thickest sediments on Earth. However, thin sediments on deep ocean basin floors imply these topographic lows were not yet present during Worldwide Flood deposition. Ocean basin subsidence is the probable cause of floodwater regression off continents.

"At the present time the oceans cover much more of the earth's surface than the continental regions. There does not appear to be a corresponding dominance of deep over shallow water sediments in the geologic column; on the contrary, ancient deep sea deposits appear to be very rare. This supports the concept of isostasy since it suggests that continents have never been submerged to great depths below the oceans" (Selley, 1970, p. 203).

God declares, *"Who shut up the sea behind doors when it burst forth from the womb ... This far you may come and no farther; here is where your proud waves halt"* (Job 38:8, 11b, NIV). Physical conditions were established at the end of the Worldwide Flood that designated a lasting relationship between modern continents and oceans.

Earth's modern mountains are regarded as far more rugged and reach much higher elevations than the Antediluvian "mountains", as presumably described by Noah's son Shem to his descendants. **Earth's Antediluvian surface probably featured land-and-lakes topography that was rapidly inundated by rising floodwaters.**

Biblical Genealogies

It is important not to read into Scripture more than Scripture provides. Abundant numeral discrepancies in genealogical records of the earliest copies of the Septuagint (Greek LXX), the Samaritan Pentateuch, and Masoretic texts yield incongruous chronological dates for Creation and the Worldwide Flood (see Young, J.A., 2003, pp. 417-430). Those who might regard the Masoritic (Hebrew) text to have been exclusively transmitted through the ages in its original God-breathed form must consider *"... the evidence that not all of the post–diluvian patriarchs are listed in our present Hebrew text. For in Luke's geneaology of Mary, the name of "Cainan" appears between "Shelah" and "Arphaxad" (Luke 3:36). The Septuagint translation of Genesis 11 places the name "Cainan" in the same position that Luke does"* (Whitcomb and Morris, 1961, p. 475). However, the Masoritic text skips "Cainan" by stating, *"When Arphaxad had lived 35 years, he became the father of Shelah"* (Genesis 11:12, NIV). In conclusion, Biblical genealogical records serve the purpose of demonstrating the purity of the Line of Christ. However, any attempt to derive specific chronological dates for Creation and the Worldwide Flood is inconclusive.

"A careful study of the Biblical evidence leads us to the conclusion that the Flood may have occurred as much as three to five thousand years before Abraham" (Whitcomb and Morris, 1961, p. 489). For example, if we assume the Worldwide Flood occurred roughly 3,000 years prior to Abram's birth at approximately 2167 B.C. (see Thiele, 1951), the Flood date would be 5167 B.C. and Creation Week about 6823 B.C. **Contradictions in Biblical genealogies infer an Earth age ranging between 6,000 and 10,000 years.** In contrast, Naturlaist assumptions regarding geology prescribe a 4.6 billion year old Earth. However, the raw geologic evidence is insufficient to discredit either Earth age interpretation.

The Mystery of Mankind's Ancient Past

Noah's sons were ship builders and some of their descendants may have sustained a seafaring culture from the Worldwide Flood to historical times. *"In fact, there is ample evidence that not long after the Deluge, the descendants of Noah carried out an extensive survey of the entire globe, mapping and charting every continent!"* (Noorbergen, 1992, p. 3). Noah's descendants migrated by boat and land to settle areas with abundant natural resources and suitable post-Flood climates.

One seafaring culture must have been exploring during the early stages of the Ice Age. *"The Oronteus Finneus Map, compiled in 1513 A.D., even shows Antarctica, long before the Ice Age icepack had fully built-up on that landmass. This map accurately locates the mountain ranges, river valleys, and bays of that landmass, which are now thousands of feet under ice and snow, and so, the obvious implication is that those intrepid ancient mariners, who sailed to and mapped Antarctica, did so during the time that the Ice Age was in its early stages ... Hapgood has published pictures of many of these medieval maps which, according to their medieval Turkish and Portuguese makers, were compilations of maps which were spirited away by the Franks from the great ancient Library at Alexandria in Egypt, to Constantinople in Turkey when that great Egyptian library, founded by Alexander the Great, was destroyed at around 600 A.D."* (Nienhuis, 2006, pp. 19-20) (also see Hapgood, 1996, p. 82.)

"The medieval Turkish cartographers wrote on their maps that their source maps which had been recovered from the Library at Alexandria were very ancient, and were based upon mathematics and astronomy. They said that these source maps were made by the Tyranean Seafish of ancient Phoenicia, and of course, there is more than bountiful evidence that the Phoenicians were sailing to many parts of the world for mining opportunities during their hey-day from around 2000 B.C. until around 500 B.C." (Nienhuis, 2006, pp. 19-20) (also see Hancock, 2002, p. 477). **Pre-Phoenician seafarers potentially date back to some of Noah's grandchildren.**

PRECAMBRIAN VS. PHANEROZOIC FLOOD

The most important prerequisite for developing a scientifically-viable geomorphic Flood model is properly identifying the portion of the Rock Record that correlates to Worldwide Flood deposits. A wrong stratigraphic choice dooms the model to awkward geomorphic interpretations and author-imposed miracles that are unable to compete with Naturalist Old Earth counterpart models in terms of scientific accountability.

God told Noah, *"I am surely going to destroy both them* (Antediluvian people) *and the earth"* (Genesis 6:13b, NIV). The Hebrew word *"shachath"* (7843) translated "destroy" above also means to batter or utterly waste". *"By these waters also the world of that time was deluged and destroyed."* (2 Peter 3:6, NIV). The Greek word *"apollumi"* (622) translated "destroyed" above means "to destroy fully". A one-time Worldwide Flood cataclysm infers geologic energy would have generated tectonic disturbance, volcanism, erosion, sediment transport, and deposition at a magnitude unequaled elsewhere in the Rock Record (Reed, et al., 1996). Therefore, it is reasonable to assume that Antediluvian evidences were mostly if not entirely obliterated by the Worldwide Flood. For "trial balloon" purposes, let us compare Precambrian to Phanerozoic stratigraphy to determine which exhibits greater catastrophism.

Worldwide Flood Deposit Thickness

Sedimentary formations are continuously being eroded and redeposited elsewhere. Therefore, it is impossible to quantify what volume of Precambrian strata has been eroded and redeposited as Phanerozoic strata. However, the thickness of sedimentary formations provides a gage corresponding to original volume of sediment deposition. The thickness of Precambrian strata has sometimes been characterized as inadequate to qualify as Worldwide Flood deposits. *"Yet the thicknesses of the Precambrian strata sequences are enormous, even greater than the cumulative thicknesses of Cambrian-Recent* (Phanerozoic) *strata. Furthermore, numerous geological processes seem to have been involved in producing these strata sequences, sometimes with several episodes of deformation followed by erosion, further deposition, and then further deformation ... there is a substantial record of Precambrian rock units that surpasses in magnitude the Cambrian-Recent rock unit sequence ... intense study of these metasedimentary terrains globally in recent decades – with modern saturation technology such as satellite imagery, airborne geophysics, deep seismic probing methods, plus exhaustive field work and many drill holes – has produced an incredible array of empirical data that has overwhelmingly confirmed and added to initial correlations; so that the physical reality of the Precambrian rock record can no longer be in question"* (Snelling, A.A., 2009, pp. 321, 323, 326-327). **The large thickness of Precambrian strata easily accommodates Worldwide Flood intrusion, metamorphism, and deposition.**

Because Precambrian strata do not contain body fossils, paleontologic correlations are lacking. Therefore, Precambrian dates are limited to more speculative radiometric decay dates corresponding to cosmology assumptions. Precambrian strata is subdivided into three lithologic segments. The upper "Proterozoic" segment is composed primarily of high-energy sedimentary formations. The next deeper "Archean" segment is composed primarily of intensely intruded, folded, fractured, and metamorphosed granite and gneiss. The deepest

"Hadean" segment of Earth's crust is presumably composed primarily of untramafic minerals.

For Worldwide Flood comparison purposes, let us presume the Archean segment corresponds to porous pre-continental crust that was crushed, extensively intruded, and metamorphically altered during the Worldwide Flood. The middle and lower Proterozoic segment corresponds to sedimentary materials eroded, transported and deposited during the Worldwide Flood. However, the uppermost Proterozoic Ediacaran strata (where present) is presumed to correlate to early post-Flood deposition.

Time of Deposition

Catastrophic Worldwide Flood year deposits should correlate to a thick segment of the Rock Record corresponding to an anomalously short timespan. Phanerozoic Flood models typically advocate the Paleozoic, Mesozoic, and in some cases Lower and Middle Cenozoic sedimentary formations as Worldwide Flood deposits. Radiometric decay dating methods are unreliable as specific indicators of geologic time. However, they do provide a somewhat uniform old-to-young sequence of dates spanning the Rock Record. Assuming radioactive decay rates have remained relatively constant through time, samples correlating to the Worldwide Flood Year should exhibit a narrow range of dates compared to non-Flood deposits. *"With respect to the range of* (K-Ar & Ar-Ar) *ages, Precambrian rocks showed a tighter range of ages that was within the range of ages of all the Phanerozoic eons"* (Overman, 2013).

In similar fashion, Grand Canyon diabase sills (lower-Proterozoic) and Cardenas Basalt (mid-Proterozoic) were rubidium-strontium isochron dated respectively at 1.07 billion years and 1.1 billion years. That dating relationship implies a similar timing of magma intrusion and volcanism, even though the diabase sills and dikes are not found in direct association with the Cardenas Basalt (see Snelling et al., 2003, pp. 269-270). **Preliminary indications are that Proterozoic strata spans a much shorter range of radiometric decay dates than Phanerozoic strata traditionally attributed to the Worldwide Flood. That favors Worldwide Flood deposits in the Proterozoic.**

Unconformity Context

The Greatest Angular Unconformity separates Archean rocks below from Proterozoic rocks above. Erosion in one area results in subsequent deposition in another area. During the Worldwide Flood Transgression (days 1-150), erosion obliterated the Antediluvian surface, forming the Greatest Angular Unconformity. During the initial 40 day inundation, rapid lateral flow generated extensive erosion of the Antediluvian surface. Lateral flow velocities slowed as floodwater deepened, causing intense erosion to subside. Deposition of eroded materials suspended in floodwaters predominated from day 41 to day 150. Tectonic activity associated with the Flood Regression off the continents (days 150-371) tilted Transgression Stage Formations. Extremely rapid flow of floodwater shed from emerging continents sheared off uplifted unconsolidated Proterozoic deposits to form the Great Unconformity.

The two most extensive unconformities in the world are the Greatest Angular Unconformity (at the base of Proterozoic sedimentary formations) and the Great Unconformity (at the top of Proterozoic sedimentary formations). These two predominant unconformities merge where Proterozoic sedimentary formations are absent. By comparison, Phanerozoic unconformities of later times are weakly distinguished and limited in aerial extent. Phanerozoic unconformities are more reasonably attributed to local and regional post-Flood events.

Proterozoic sedimentary formations are commonly tilted at a steeply inclined angle and erosionally truncated, indicating substantial tectonic movement interceded between Proterozoic and Phanerozoic deposition processes. The "Greatest Angular Unconformity" is most reasonably identified as a Worldwide Flood transgression feature. The "Great Unconformity" above is most reasonably identified as a Worldwide Flood Regression and very early post-Flood erosion feature. **Proterozoic sedimentary formations sandwiched between these two predominant unconformities are stratigraphically favored as Worldwide Flood deposits.**

Metal Ore Lithologic Changes

The magnitude of Worldwide Flood tectonics would probably exhibit geomorphic characteristics unlike post-Flood tectonics. Magmatic, volcanic, and hydrothermal processes generating unique metal ore deposits abruptly changed at the Precambrian/Phanerozoic boundary. *"... there was significant volcanic activity with large hydrothermal systems active during the time these upper Archaean rock sequences were formed, but there was also active erosion to produce conglomerates that hosted the gold and uranium minerals ... the volcanic activity and hydrothermal systems responsible for the volcanic-hosted massive sulfide Abitibi-type ore types and the gold and base metal veins appear to have abruptly ceased ... Suddenly, there is a cessation of banded iron formations ... At the Precambrian-Cambrian boundary there is a dramatic change, with many new styles of ore deposits occurring in the Cambrian-Recent* (Phanerozoic) *rock sequences ... Copper, molybdenum, and tin-tungsten ores are hosted by numerous granitic intrusions ... The other types of volcanogenic massive sulfide deposits are dispersed periodically through Paleozoic strata ... Another class of sediment-hosted massive sulfide ore deposits that is almost exclusively found in Cambrian-Recent strata sequences are lead-zinc sulfide ores in carbonate rock units"* (Snelling, A.A., 2009, pp. 361-364). **The Precambrian to Cambrian transition of metal ore intrusive processes of different character signals a change from higher-intensity Worldwide Flood tectonics to intermittent post-Flood tectonics.**

Precambrian Stratigraphy

Extreme catastrophism is best exhibited by Precambrian stratigraphy. *"With fracturing of the crust and release of its waters, decompression effects would have been significant, and hot rocks at depth would have melted and flowed upwards. This is envisaged as an avalanche process. Since the surface was covered with water, this would have resulted in pillow lavas, peperites and hyaloclastic deposits, in addition to extensive igneous activity and massive sedimentation. All the sediments coming from the antediluvian world would be mingled with the products of igneous activity, accompanied by extensive metamorphism, and this had the effect of obliterating anything that might potentially have become a fossil ... For* *more on peperites, see Walker (2004), noting that these textures are widely documented in the Precambrian"* (Tyler, 2006, p. 76).

Laminated cyclic and rhythmic deposition exhibited by the 18 km thick Precambrian Belt Supergroup of Montana, U.S.A. implies a continuous depositional process, distinguished by flow-induced grain-size sorting. The relative lack of distinct sequence (formation and member) boundaries plus the absence of burrowing structures implies relatively continuous deposition of Precambrian Belt Supergroup strata from bottom to top. Sedimentary structures are characteristically marine, with no evidence of windblown sand. Belt Supergroup sediments exhibit an abundance of exceptionally fine-grained materials, with few grains coarser than sand size. That implies extremely abrasive erosion and transport conditions that demolished full-bodied lifeforms as they bounced along the bottom. If the stromatolite features in the upper portion of the Belt Supergroup are organic, they would represent a continuous depositional transition spanning late Worldwide Flood Regression into earliest post-Flood times.

Floating microorganisms had the best chance of surviving the Worldwide Flood intact as recognizable organisms prior to being buried and fossilized. Denser-than-water hydrosulfuric acid would have been most concentrated adjacent to the seafloor. Acidic dissolution of lifeforms that sank to the seafloor would have proceeded rapidly.

Heat decomposition of "floaters" would have been minimized at the surface by evaporative cooling. "Floaters" also avoid most mechanical abrasion with solid surfaces and turbulently suspended particles. Capillary attraction of water clinging to microrganisms and tiny particles acts as a cushion, retarding pulverization. In accordance, microfossil acritarchs such as unidentified spores, pollen, algae, phytoplankton, and dinoflagellates are found in some Proterozoic Grand Canyon Supergroup formations. In contrast, sinking Antediluvian biogenic materials were probably pulverized and dissolutioned into unrecognizable organic materials prior to burial. "Molar Tooth" structures are considered to have been generated from bacterial decomposition degassing of pulverized organic materials in Proterozoic carbonates.

Worldwide Flood sediments were initially unconsolidated and therefore were vulnerable to being eroded and redeposited multiple times. Some Transgression Stage deposits were stabized by intruding dikes by and metamorphosing fluids. Most remnant Proterozoic sedimentary formations are Transgression Stage deposits that were preserved in topographic depressions not subject to Regression Stage and younger erosion. Uppermost Proterozoic "cover" corresponds to remnant clasts from eroded Flood deposits plus very early post-Flood erosion and deposition. Continental shelf deposits were primarily sourced by materials eroded from continents during the Regression Stage and soon after the Worldwide Flood.

Post-Flood Erosion and Deposition

Early post-Flood land would have featured negligible vegetation to stabilize the land surface and inhibit erosion. It is reasonable to assume that massive erosion and deposition would have occurred during the years following the Worldwide Flood. Rapid erosion would perpetuate denudation of the post-Flood landscape.

Unconsolidated sediments would have been saturated following the Worldwide Flood. *"Thus, thick post-Flood deposits would be expected from massive post-Flood erosion, regardless of where one places the post-Flood boundary ... Water saturated rocks and sediments have been one of the key factors in instigating many types of mass movements (landslides, mudflows, slumps, etc.) ... It is unknown how long it took connate Flood water to sufficiently drain out of Flood sediments; but potentially this could have been an important factor for centuries or millennia. ... Water saturated sediments could have been more easily eroded by rivers that began to flow after the Flood because of groundwater sapping ... As rivers (and other processes) cut deeper into the post-Flood landscapes, sediment and rock filled with connate water could have easily been removed by sapping ... Sapping would not only have dramatically widened post-Flood valleys, it would have greatly lengthened them as well"* (Whitmore, 2013, pp. 3-4).

The Fountains of the Great Deep and tectonic activity would have produced hot post-Flood oceans. Consequent high rates of evaporation would have generated prolific rainfall for centuries or millennia before stabilizing at cooler modern temperatures. *"If large storms and hypoercanes developed after the Flood, they would have had significant erosional and depositional consequences"* (Whitmore, 2013, p. 4). High precipitation rates would have accelerated mineral decomposition rates, rendering materials more vulnerable to being eroded.

Prolific earthquake-induced fault displacement ranging from local to mountain-building in scope would have steepened slopes, destabilized materials, increased surface flow velocities, and accelerated rates of erosion. *"It is thought that earthquake intensity and frequency has been exponentially decreasing since the time of the Flood (Austin, 2010), thus mass wasting events were likely larger and more common as a result of these processes in the years immediately following the Flood than they are today"* (Whitmore, 2013, p. 7).

"Geomorphic theory suggests rivers tend to erode (or deposit) until they reach a "graded" profile as long as they are flowing on "adjustable" materials (Bloom, 1998) ... The further they are from grade the higher the rates are. In immediate post-Flood times one cound imagine that landscapes were significantly "out of grade" because of recent tectonic uplift to raise the continents from the Flood waters. Additionally, poorly consolidated sediments could be considered to be very "adjustable." Considerable erosion (and deposition by rivers could be expected until graded profiles were reached. Additionally, tectonic activity and isostatic adjustment would probably continue to radically change river profiles, throwing them out of grade, leading to further erosion and deposition. Combined with higher post-Flood precipitation rates and drainage from water-saturated sediments, rivers would have been much larger than those we find today" (Whitmore, 2013, p. 8).

"Because of expected post-Flood erosion that would occur, irrespective of where one places the post-Flood boundary, enormous quantities of sediment should be found resting on the post-Flood unconformity" (Whitmore, 2013, p. 15**). In conclusion, the Cambrian through Pleistocene portion of the Rock Record may reasonably be attributed to rapid erosion/deposition processes spanning several centuries following the Worldwide Flood.** Recent sedimentary formations exhibiting a much lower frequency of catastrophic erosion/deposition features correspond to stabilized geomorphic processes spanning the millennia that followed.

Sinking Continental Basins

Continental basin sedimentary formations are typically tens of meters thick in the basin center, thinning outward to a feather edge along the basin rim. Central basin thickening indicates continuing subsidence during deposition. Sedimentary formations were deposited on a sloping basin surface, dipping gently toward the sinking basin center. The mixing of Paleozoic sediments and fossils transported from distant ecosystem source areas is minimal. That implies a marine habitat located far from land or adjacent to land that is too flat to erode and transport clastic sediments offshore.

Paleozoic carbonate formations appear to have been precipitated in place during many separate deposition events. Paleozoic formations also exhibit distinctive fossil ecosystem assembledges. However, there was insufficient time to repeatedly recolonize, destroy, bury, and preserve these marine ecosystems during the Worldwide Flood.

Continental basins usually exhibit a Paleozoic sequence of marine formations, with each higher formation displaying progressively shallower water ecosystem fossils and lithology. Coexisting, concentric ecosystem rings progressively migrated toward the basin center. Each ecosystem ring overlapped and deposited its shallower water fossils and lithology on top of its deeper water neighbor. Deeper water marine ecosystems were systematically eliminated at the basin center as they were blanketed by the lateral advance of each shallower ecosystem neighbor.

Mesozoic formations exhibit an increasing upward abundance of terrestrial fossils representative of unique subaerial ecosystems. Swamps transition upward into lowland ecosystem formations. Distinctive Mesozoic formation boundaries are characteristic of lateral sediment transport restricted to a few kilometers or less. Cenozoic formations continue this overlapping stratigraphic stacking of successively higher elevation ecosystem features. Continental basin subsidence was routinely being overwhelmed by sediment infilling. **The Phanerozoic stratigraphic sequence corresponds to regressing continental basin seas being progressively displaced by land.**

Could most of the Phanerozoic stratigraphic sequence correspond to Worldwide Flood deposition? An Inundation Stage burial sequence beginning with deep water marine and ending with highland ecosystem evidences would potentially accommodate the observed the Phanerozoic strata sequence. However, stratigraphic context for the Worldwide Flood deposition of most Phanerozoic strata is problematic. For example, plants provide the greatest capacity for depositional segregation by flotation. However, the absence of anticipated hydrodynamic sorting is confirmed by the failure to segregate variable-density arborescent plants from herbaceous plants. Also, pitch-laden gymnosperms should float longer than angiosperms, but the opposite order is observed in the fossil record (Wise, 2003, p. 374).

Body Fossil Transport

By comparison, Phanerozoic (Cambrian-Recent) stratigraphy characteristically exhibits much lower intensity catastrophic evidences than Precambrian stratigraphy. Paleozoic and Mesozoic sedimentary formations commonly contain abundant body fossils inferring short-distance transport or in-place burial. In contrast, intense Worldwide Flood erosion and long distance cataclysmic transport would be expected to pulverize body fossils prior to deposition.

The sedimentary layer-cake of Phanerozoic formations in continental basins commonly extends for hundreds of kilometers in diameter. At any specific location, each sedimentary layer must be deposited before the next sedimentary layer above is deposited to avoid mixing and to preserve distinctive lithologic characteristics. Long-distance transport causes materials to be sorted by density and shape. However, Phanerozoic formations typically exhibit a characteristic fossil assembledge, containing high-density, low-density, durable, and fragile fossils together. That stratigraphic evidence is indicative of in place burial or very short-distance transport preceeding deposition. Distinct formation lithologies separated by abrupt formation boundaries are characteristic of unique material sources and depositional processes, which seem impossible to achieve by any reasonable long distance transport mechanism.

Short-distance Worldwide Flood body fossil entombment as described above would be possible for a basal Worldwide Flood formation. However, all overlying formations would require a seemingly unviable long-distance hydrodynamic transport flow mechanism. The Worldwide Flood source area for Phanerozoic body fossils must be located outside of the area already blanketed by underlying formations and outside the area from which all underlying sedimentary formations were previously eroded and transported. Material transport distances would extend to hundreds of kilometers after two or three sedimentary formations were deposited.

Individual Paleozoic carbonate formations in continental basins typically exhibit a "bulls-eye" pattern of deep water lithologies and fossil assembledges which extend from basin center outward to progressively more shallow lithologies and fossil assembledges on continental basin rims. That pattern is indicative of established ecosystems and lithologies deposited in place. On a smaller scale, individual Paleozoic formations commonly exhibit multiple marine transgressions and regressions (designated "members"), which are traced back and forth, by following adjacent laterally-shifting ecosystem deposits vertically through the Rock Record. Yet these apparently shifting paleo-ecosystems commonly rest above hundreds of meters of other autonomous paleo-ecosystem deposits. Stratigraphic context requires specifying a reasonable cause-and-effect mechanism to explain that unexpected erosion-transport-deposition process.

Body fossils pulverized by catastrophic Precambrian erosion and transport would not have been available for subsequent Phanerozoic Flood burial. Where were those Phanerozoic materials during the preceeding high-energy Precambrian deposition? How were those Phanerozoic materials transported from afar and deposited as lithologically and paleontologically segregated formations that so closely resemble modern ecosystem relationships? **Making a convincing stratigraphic context argument advocating a large segment of the Phanerozoic to be Worldwide Flood deposits would be difficult.**

Post-Flood Recolonization

Marine life, surviving in floodwaters outside the Ark, repopulated most rapidly. Recolonization began at the base of the food chain and worked upward. Animals higher on the food chain struggled to survive in isolated pockets until more accommodating habitats were established (Tyler, 2006, p. 80). Terrestrial Ark voyagers that *"had the breath of life in its nostrils"* (Gen. 7:22, NIV) required centuries to spread, repopulate post-Flood continents, and deposit their fossils in numbers sufficient to be preserved and found in the Phanerozoic Rock Record.

Worldwide Flood erosion denuded the emerging continental landscape. Rapid erosion of unconsolidated Flood sediments continued during the post-Flood years. The sudden proliferation of body fossils in Cambrian sediments is a convincing stratigraphic indicator of early post-Flood recolonization and rapid burial. During the Post-Flood Millennium (1,000 year timespan), unconsolidated Flood deposits were rapidly eroded, transported and deposited elsewhere in Phanerozoic strata. Abundant transported materials facilitated rapid burial of organisms in early post-Flood habitats.

Alternating phases of slow deposition and fast deposition adds up to a slow time frame. The interbedding of slow deposition layers with catastrophic layers undermines the concept that both types of layers were deposited during the Worldwide Flood Year. Intermittent periods of apparent slow stable Phanerozoic deposition are revealed by fossilized reefs, worm burrow zones, vast chalk deposits (Tyler 1996, pp. 107-113), subaerial lava flows (Garner 1996, pp. 114-127), and buried soil zones with root structures (Robinson, S., 1996, pp. 32-69). **A post-Flood Phanerozoic recolonization sequence reasonably explains the observed progression of marine life at the base of the Rock Record, transitioning upward to include more complex terrestrial lifeforms at the top.**

Fossil Reefs

Modern reefs require centuries to form. Fossil reefs are composed of sedentary marine organisms intermingled with the fossils of mobile creatures living on the reef, plus assorted fine-grained sediments. Modern reef frameworks are built by corals, calcareous hydrocorallines, rudistids, archaeocyathines, bryozoa, algae, etc. in shallow, clear, tropical seas. Worldwide floodwaters would have been too short-lived and too murky to facilitate reef growth.

The beautiful coral reef "forests" we see in underwater pictures are constructed on an accumulation of broken pieces from past storms. Reef frameworks are much too fragile to be transported by floodwaters and re-deposited elsewhere. Antediluvian reefs would have been destroyed and scattered by cataclysmic action during the Worldwide Flood. Paleoreefs provide a non-Flood indicator because time was too short and cataclysmic conditions too adverse for reef construction during the Worldwide Flood Year. However, fossil reefs are described in every Phanerozoic period of geologic history (Selley, 1970, p. 173; Heckel, 1974, pp. 90-154). **The presence of paleoreefs inhibits attributing large segments of the Phanerozoic Rock Record to Worldwide Flood deposition.**

Halite Formations

Can halite (NaCl) table salt formations blanketing large portions of continental basins be attributed to Worldwide Flood deposition? If so, how were they deposited as tabular beds within the Phanerozoic Rock Record? The relative purity of halite formations eliminates the possibility of direct igneous injection into preexisting permeable tabular-shaped sedimentary formations.

Hydrothermal fluids expelled through Earth's crust by the Fountains of the Great Deep during the Transgressive Stage of the Worldwide Flood potentially provide an abundant source of salts and sufficient heat to accelerate evaportation and associated salt precipitation. However, there is not enough time for thick salt beds to form solely by evaporation during the Worldwide Flood year. Heating to a boil would agitate the seawater, increase solubility, effectively dissolve, and maintain halite in solution.

Rapid consolidation of a halite slurry mush into rock salt seems unviable under aquatic conditions. For example, the direct precipitation of solid halite would theoretically require supercritical seawater temperatures above 430 degrees C, below the 800 degree C melting point of halite, and pressures corresponding to approximately 2800 meters in depth (Hovland, et al., 2006, pp. 855-869). Therefore, let us assume brines in that supercritical temperature range were vented at ocean basin depths (2800+ meters) by the Fountains of the Great Deep. Seawater would vaporize upon contact with the hot brines, rapidly reducing the temperature to subcritical levels. Salt precipitating out of boiling brines would either dissolve in seawater or form a dense halite slurry mush that would flow laterally along the seafloor and settle in topographic depressions. The point being supercritical temperatures could not be maintained in an unbounded aqueous environment for long enough to consolidate halite into solid rock. Consequently, the Worldwide Flood did not provide conditions facilitating deposition of tabular halite sedimentary formations as observed in the Phanerozoic Rock Record. (Any directly precipitated solid halite would probably form a chimney or volcanic cone surrounding the vent instead of the observed tabular formation shape.)

Compare the density of halite (specific gravity 2.16) to other common low density minerals such as quartz (2.65), orthoclase (2.57), plagioclase (2.62-2.76), muscovite (2.76-3.1), and biotite (2.8-3.2). If higher density unconsolidated sediments were deposited on top of halite, the halite slurry mush would flow rapidly upward to the seafloor surface and be dispersed. Clastis minerals raining down from seawater would sink to the base of the halite slurry. That process would have prevented entrapment of tabular halite formations beneath other unconsolidated sediments. Halite slurries in floodwater would have relentlessly risen to occupy topographic surface depressions above all other Worldwide Flood deposits. Therefore, it is reasonable to expect rock salt to appear in the Rock Record at the top of Worldwide Flood deposits and in subsequent post-Flood subaerial deposits.

The segregation of different mineral salts implies shallow evaporative precipitation affected by varying chemical and temperature conditions in accordance with relative solubility. There is a well-known precipitation sequence that comes from shallow sea water evaporation in arid climates, starting with calcite (2.72), then gypsum (2.32), then halite (2.16), then sylvite (1.99) (Boggs, 1986, p. 193) which is commonly observed in tabular mineral salt formations. Thin tabular salt formations exhibiting this vertical deposition sequence are most reasonably interpreted as post-Flood evaporative precipitates.

However, thick tabular halite formations may also be deposited directly by magma extrusion on subaerial surfaces. For example, thick pure halite formations suggest halite magma segregated from other mineral salts at depth before being extruded. The melting point of halite (NaCl) is 801 degrees C. The melting point of anhydrite (CaSO4) is 1450 degrees C. At temperatures between 801 and 1450 degrees C, anhydrite (2.89-2.98) would settle out as a solid below liquid halite (2.16) magma. Optimum conditions for magmatic extrusion, deposition, and consolidation of thick tabular halite formations follow:

1.) Playa setting – a relatively flat surface in an arid climate with no rivers, streams, or surface drainage features.
2.) Low-density, low-viscosity halite-rich magma sourced from Earth's interior segregating from other minerals and accumulating at depth.
3.) Overlying materials blocking halite magma rise to the surface are isostatically lifted by accumulating halite-rich magma, fracturing that impermeable barrier above.
4.) Halite-rick magma rises through fractures and is rapidly extruded onto the playa surface.
5.) Vacating the halite magma chamber below causes the surface above to subside.

6.) Surface subsidence forms a depression containing the extruded halite-rich magma. (The magma is being transferred from a subsurface position to a surface position directly above.)

7.) Water (1.00) contained in the halite magma rises to the top of the extruded halite (2.16), and evaporates, preserving the relative purity of the halite deposit.

8.) Dewatering also produces a gentle topographic surface depression, preventing surface water dissolution and transport removal.

9.) Dust storms between separate halite extrusion events form recognizable markers.

10.) Clastic sediments eroded and transported from outside the depression blanket the halite formation in place.

11.) Accumulating clastic sediment overburden pressure further dewaters and consolidates the halite into a crystalline rock salt formation.

12.) The preceeding process may be repeated multiple times if subsurface fractures in the strata below temporarily reseal and the magma chamber is progressively refilled with halite-rich magma.

13.) The rock salt formation may subsequently be buried to sufficient depth to increase pressure, causing the rock salt to flow laterally toward subsurface highs in the capping formation above.

14.) The low density of accumulating salt causes vertical uplift and fracturing of the overlying cap rock.

15.) Halite rises up the fractures in the form of a salt dome.

The Collapse Tectonics Model timeline follows:

1.) Worldwide Flood erosion bevels emerging continents.

2.) During the Post-Flood Millennium, some continental interior playa areas experience large-volume subaerial extrusion of halite-rich magma.

3.) Surface formations subside into the subsurface void vacated by escaping halite-rich magma.

4.) Clastic sediments eroded and transported from adjacent higher topographic features blanket, overpressure, and further dewater the halite, facilitating the consolidation of halite.

5.) Mountain upheaval 800 to 1,000 years after the Worldwide Flood raises mountain ranges in adjacent areas. Erosion, sediment transport and deposition increase overburded depth to complete the dewatering and consolidation of halite into crystalline rock salt at depth.

6.) During the millennia that follow, various locations experience salt dome flow, overburden strata removal by erosion, and/or groundwater solutioning and subsurface halite removal, causing overlying strata to collapse.

Finally, the high solubility of mineral salts requires special conditions to prevent dissolution in sea water. Without rapid consolidation into rock and protection from subsequent dissolution, tabular salt formation deposits could not be attributed to Worldwide Flood deposition. Tabular halite rock salt formations distributed throughout Phanerozoic sedimentary layers are best explained by various subaerial post-Flood deposition processes.

The deepest stratigraphic position of tabular halite deposits is commonly documented adjacent to the Proterozoic/Phaneorzoic boundary. The Collapse Tectonics Model attributes this stratigraphic interval to very early post-Flood deposition. Retreating brine water was trapped in topographic depressions as floodwater regressed off continents. Evaporation of these pools, lakes, and shallow seas precipitated mineral salts in these Proterozoic/Phanerozoic boundary formations.

Tabular halite formations spanning Phanerozoic strata provide convincing physical and geochemical evidence supporting a post-Flood deposition setting. By deduction, the bulk of Proterozoic sedimentary formations are strongly favored as the optimum stratigraphic position of Worldwide Flood deposits.

Air-breathing Land Animals

Footprints of land animals crossing mudflats are typically preserved when volcanic ash-falls occur slightly before, during, or shortly after the tracks were made. The volcanic ash quickly hardens fresh tracks to preserve fossil footprints. In one 177-meter stratigraphic section in the St. Mary Formation logged a vertical succession of more than 100 track-bearing strata (Currie et al.. 1991, pp. 102-115). Many of the beds are rooted (Nadon, 1993, pp. 31-44). Rooted beds are indicative of prolonged subaerial exposure.

Dinosaur egg nests are sometimes preserved in a vertically-stacked succession of Phanerozoic layers. Five levels are preserved in a 100-meter section at Ukhaa Tolgod in the Cretaceous of Mongolia (Dashzeveg, et al., 1995, pp. 446-449). At least six levels are preserved in the Two Medicine Formation (Horner and Gorman, 1988). Five levels are preserved in a 100-meter section near Aix-en-Province, France (Cousin, et al., 1994, pp. 56-74). Dinosaur nests and footprints (Garton, 1996, pp. 82-106) verify land animal habitation. Dinosaur nests with hatched eggs indicate a prolonged subaerial depositional setting not easily explained during Worldwide Flood conditions.

Bones may be transported, but footprints in rooted soils and nests with eggs affirm dinosaur habitation. *"Most reptiles bury their eggs in the ground or place vegetation on top. In this way the eggs stay in an environment of controlled high humidity, high CO_2, and low O_2 ... an examination of dinosaur eggs has shown that the egg shells are very porous, generally like those of reptiles"* (Oard, 2013, p. 3). The ideal habitation for dinosaur egg-laying would have been adjacent to swamps and intertidal areas along the flanks of continental interior seas. These low-lying, post-Flood terrestrial ecosystems provided optimum conditions for rapid sediment deposition. Dinosaur eggs and footprints were preserved and fossilized in place.

"A scenario which has hosts of dinosaurs, reptiles and mammals during the Flood swimming around, making tracks, laying eggs, and digging burrows, above kilometers of volcanics, carbonates and clastics deposited earlier in the Flood while they were somewhere else, would be difficult to believe in" (Robinson, S., 2003, *TJ*, 17(3), p. 55). **The dinosaurs were descendants of Ark voyagers – not Antediluvian animals.**

Post-Flood Calamities

"The sea looked and fled, the Jordan (River) *turned back;*
The mountains skipped like rams, the hills like lambs"
(Psalms 114: 3-4, NIV).

The tectonic activity described above during the Exodus is of comparable magnitude to intermittent regional tectonic events being proposed herein to have occurred during the early post-Flood centuries. By comparison, these regional post-Flood calamities would have been orders-of-magnitude smaller than the Worldwide Flood cataclysm. However, most of these regional catastrophic events would have occurred in continental interiors. In contrast, continental rim areas where human habitation was concentrated were comparatively stable and sufficiently distant avoid frequent continental basin calamities. Humans avoided taking up residence in muggy, insect-infested, dinosaur-occupied lowlands close to subsiding continental basin seas, where tsunamis obliterated adjacent areas every decade or so. Post-Flood mountain upheaval was also life-threatening to man and beast. However, there were always plenty of habitable post-Flood areas where man and beast could survive.

Assuming the Worldwide Flood was accompanied by an extensive disruption of the Antediluvian crust, tectonic activity would not have abruptly halted. For example, aftershocks follow earthquakes as Earth's crust stabilizes. On a much grander scale, catastrophic tectonics associated with the Worldwide Flood probably persisted at a declining rate for centuries thereafter.

One post-Flood collapse of Earth's crust is referred to in the account of Korah, Dathan, Abiram and their followers. *"As soon as he* (Moses) *finished saying all this, the ground under them split open and the earth opened its mouth and swallowed them, with their households and all their possessions. They went down alive into the grave, with everything they owned; the earth closed over them, and they perished and were gone from the community"* (Numbers 16:31-33, NIV).

One historical example of a collapse event occurred at the City of Euphemia in Calabria in the "ball of the foot" of southern Italy in 1638. *"Here scenes of ruin everywhere appeared around me; but my attention was quickly turned from the more remote to contiguous danger, by a deep rumbling sound, which every moment grew louder. The place where we stood shook most dreadfully; after some time, the violent paroxysm ceasing, I stood up, and turning my eyes to look for Euphemia, saw only a frightful black cloud. We waited till it had passed away, when nothing but a dismal and putrid lake was to be seen where the city once stood"* (Kircher, 1828, p. 75).

A modern example of catastrophic crust collapse occurred during August of 1999, when a six square mile area in the Turkish port city of Izmit collapsed below sea level, causing 13,000 deaths. *"Only the tops of an amusement park Ferris wheel and roller coaster remained visible above the waves"* (T. Budd, 2000, relief effort eyewitness account). God's covenant with Noah states, *"Never again will all life be cut off by the waters of a flood; never again will there be a flood to destroy the earth"* (Genesis 9:11b, NIV). **God's covenant promising not to bring a second Worldwide Flood remains intact because post-Flood inundations were restricted to local and regional areas.**

COLLAPSE TECTONICS MODEL

Geomorphic models of Earth's past cannot be proven because they are one-time unrepeatable events. Model evaluations are similar to circumstantial cases in a court of law. A cause-and-effect mechanism must be identified that reasonably explains the geologic evidence. In this case, the cause is the tectonic driving mechanism and the result is observed geologic evidences. However, without a scientifically-accountable driving mechanism, the relationship between cause and effect is broken. (For example, the Plate Tectonics mantle convection cell driving mechanism is challenged in Addendum 1.) Collapse Tectonics counters with porous crust collapse and continental root underplating as scientifically-accountable driving mechanisms. So put on your judicial robe and objectively decide if Collapse Tectonics provides a more reasonable driving mechanism explaining the observed geologic evidences.

The following timetable correlates Scriptural and archaeological records with Collapse Tectonics Model parameters. **An Earth age of 6,000 to 10,000 years is sufficient to accomplish all of the geologic work necessary to generate Earth's modern landforms via Collapse Tectonics processes.**

COLLAPSE TECTONICS MODEL CHRONOLOGY

LXX TIMELINE	OCEANS	CLIMATE	CONTINENTS
^ ^ ^			
0 AD +/- (Jesus born)	Cold	Dry & Sunny	Subdued Tectonic
1450 BC +/- (Exodus)	Oceans	Atmosphere	Catastrophism
2000 BC +/- (Abraham)			
2200 BC +/- --			Pleistocene Ecosystem Adaptation & Burial
	Warm Oceans	Humid & Cold Atmosphere	Ice Age Glaciation
2400 BC +/-			Mountain Range Upheaval
	Intermittent Sinking of Mid-Atlantic Ridge		Cenozoic Ecosystem Recolonization & Burial
POST-FLOOD MILLENNIUM			Mesozoic Ecosystem Recolonization & Burial
Time of Intermittent Catastrophic Collapse of Porous Continental Basin Crust	Hot Oceans	Hot Atmosphere	Paleozoic Ecosystem Recolonization & Burial
	Mid-Atlantic Ridge Upheaval		Perched Ediacaran Seas on Denuded Land Surface
3201 BC +/- --			Ark Departure – Day 370
	Rapid Continental Shelf Deposition	Windy Albedo-driven Weather Patterns	**Great Unconformity**
			Continental Rim Underthrust/Upheaval
Regresssion	*Catastrophic Collapse of Porous Ocean Basin Crust*		Land Exposed – Day 224
			Ark Grounded – Day 150
WORLDWIDE FLOOD	Fountain Flow Ceases		
			Min. Depth 20' – Day 40
Inundation			*Proterozoic Deposits*
	Ocean Trenches Eroding		**Greatest Unconformity**
3200 BC +/- ----- *Fountains of Great Deep Erupt* -------------------------------- – Day 1			
ANTEDILUVIAN AGE	Pyromagma Solidification of Porous Crust/**Continental Root Uplift**		
	Worldwide Land & Lakes Topography w/ Primordial Ecosystems		
4856 BC +/-	*------------- **SIX DAY CREATION** -------------*		

Magma Types

Primordial Earth experienced progressive density and viscosity segregation of magma types. Magma components naturally segregate according to differing physical characteristics. The magma types listed below are advocated as idealized end products, which grade into each other. For descriptive purposes, it is useful to identify four typical categories of magma by density stratification, arrival at the base of Earth's crust, solidification sequence, primary mineral-forming components, and (specific gravity):

1.) *Pyromagma* – low-density, alkaline, low-viscosity, extremely low-silicate magma rose rapidly to the base of Earth's primordial crust. During the Antediluvian Age, *pyromagma* was the primary crust-forming material. Contains abundant volatile components of water (1.0) and carbon dioxide. Modern pyroclastic eruptions expell volcanic ash and fine-grained materials instead of lava. Primary mineral solidification forms halite (2.16), gypsum (2.32), calcite (2.72), dolomite (2.85), and apatite (3.15-3.2).

2.) *Hypomagma* – low-density, acidic, viscous (slow flowing) high silicate magma with moderate water content. Typically solidifies at lower temperature than other magma groups. Segregated deep in Earth's interior and rose slowly up continental root conduits, then mushroomed outward to form a deep under-plate below pre-continental Antediluvian crust. Formed much of Earth's modern continental crust. Primary mineral solidification forms orthoclase (2.57) and quartz (2.65).

3.) *Basaltic magma* – medium-density, alkaline, medium-viscosity silicate magma with low water content. Segregates below magmas listed above. Solidified primarily after being injected into the overlying collapsed oceanic crust, or as extruded lava flows during the Worldwide Flood Regression and during post-Flood mountain-building at the close of the Post-Flood Millennium. Primary mineral solidification forms calcic plagioclase (2.76).

4.) *Ultramafic magma* – high-density, alkaline, viscous silicate magma with the lowest water content. Formed the majority of earth's upper mantle beneath the magmas listed above. Occasionally extruded during post-Antediluvian tectonic activity. Primary mineral solidification forms pyroxene minerals (typically 3.2-3.5) and olivine (3.27-4.37).

A single volcanic eruption may extrude any combination of the four magma types listed above. *Hypomagma, basaltic magma,* and *ultramafic magma,* are all silica-rich. In contrast, *pyromagma* components exhibit low viscosity and separate more readily from viscous, silica-rich magmas. All magmas exhibit ductile or liquid flow and segregate by specific density stratification. For example, granitic intrusives commonly transition downward into higher density migmatites.

Collapse Tectonics proposes the Antediluvian crust solidified from the surface downward, encapsulating liquid-filled vugs containing *pyroclastic* volatiles (liquids). Three potentially viable conditions are required to validate that concept:

Condition 1.) The low-density and low-viscosity of *pyromagma* facilitated its rapid rise and early concentration at the base of the primordial crust to become the primary Antediluvian crust-forming magma.

Condition 2.) The primordial Antediluvian crust acted as an unbroken seal, preventing *pyromagma* volatiles from escaping upward to the surface. That impermeable pre-Flood barrier was continuously maintained between the base of the downward solidifying crust and deepening lineament fractures during the Antediluvian Age.

Condition 3.) *Pyroclastic* volatiles became supersaturated in the crust-forming magma zone, but were too low-density to be displaced downward into higher-density magma components below. **Therefore, low-density *pyromagma* volatile components were encapsulated as liquid-filled vugs within the downward-solidifying crust to form a porous Antediluvian crust.**

COLLAPSE TECTONICS SUMMARY

The Collapse Tectonics Model is constructed upon two foundational **Postulates (P)** that lead to thirty-five **Derived Conclusions (DC).**

P 1 – Earth's primordial crust formed a tectonically-stable permeability barrier approximately 2 km in thickness, preventing upward migration of volatiles. Low-density, *"pyromagma"* segregated and rose rapidly from Earth's interior to the base of the primordial crust to form a crystal melt mush zone. The crystal melt mush solidified downward during the Antediluvian Age to depths ranging between 20 and 25 kilometers. Volatile components were expelled into the magma melt below as Earth's crust cooled and solidified downward. Those volatile fluids soon became over-saturated in the magma melt, but were too low density to be forced downward. Tiny "vugs" (millimeters in diameter) containing liquid volatiles were encapsulated as one-third of rock volume within a predominantly pyroclastic rock matrix.

P 2 – During the Antediluvian Age and thereafter, viscous, low-density, quartz-rich *hypomagma* segregated deep in Earth's interior and rose slowly in vertical columns known as "continental roots". Ascending *hypomagma* bumped into the crustal ceiling, causing the crust above to bulge upward. Hypomagma mushroomed outward, under-plating Earth's pre-continental crust. The spreading *hypomagma* underplate flowed around "bowl-shaped" lenses of low-density *pyromagma* and displaced *basaltic magma* laterally into areas beneath pre-oceanic crust. Deep *Pyromagma* pockets solidified downward during the Middle and Late Antediluvian Age, producing regions underlain by anomalously thick pre-continental basin porous crust. Thirty percent of Earth's Late Antediluvian crust was underplated by *hypomagma*, which would form future continents. Approximately seventy percent of Earth's surface (a.k.a. pre-oceanic crust) was characterized by the absence of a *hypomagma* underplate.

DC 1 – Continental roots progressively sourced *hypomagma* that mushroomed outward, underplating the base of the pre-continental Antediluvian crust. Isostatic uplift raised the pre-continental crust above in accordance with the increasing thickness of the spreading, low-density *hypomagma* underplate.

DC 2 – Isostatically-uplifting pre-continental crust generated tension in pre-oceanic areas paralleling the pre-continental/pre-oceanic crust boundary. The Worldwide Flood was initiated when normal faulting extended downward, penetrating the crust-forming magma zone and rupturing the previously impermeable Antediluvian crust.

DC 3 – Water-rich *pyromagma* volatiles flowed laterally beneath the pre-oceanic crust and upward through faulted lineament conduits to be expelled as "Fountains of the Great Deep". The rapid upward escape of overpressured *pyromagma* from the crust-forming magma zone eroded deep oceanic trenches.

DC 4 – Water-rich volatiles extruded by the "Fountains of the Great Deep" sourced torrential rains. The resulting downpour caused cataclysmic surface flow and erosion, carving the Greatest Angular Unconformity on shallow Antediluvian surfaces.

DC 5 – Inundation-stage erosion was greatest across pre-continental topographic highs. Erosion removed the primordial crust from those topographic highs, initiating isostatic uplift, tectonic fracturing, porous crust collapse, magma intrusion, and metamorphic alteration. Magma intruded collapsing porous crust, forming Archean granite-gneiss complexes. Lava was extruded as submarine volcanic flows. Intense volcanic activity formed "greenstone belts" above major fracture zones.

DC 6 – Two tidal swells (one facing and one opposing the orbiting moon) rotated around the circumference of the globe. Tidal waves rose across shallows, beveling topographic highs.

DC 7 – Resulting sediment transport and deposition infilled adjacent pre-continental topographic lows with Proterozoic sedimentary formations and began to accumulate along shallow pre-continental margins.

DC 8 - Transgressing floodwaters completely inundated Earth's gently-rolling land-and-lakes surface to a minimum depth greater than six meters.

DC 9 – Pyromagma escaping from the pre-oceanic crust-forming magma zone caused the overlying seafloor to settle and fracture. Contact of porous crust with underlying hotter *basaltic magma* induced porous crust melting and accelerated volatile escape through deep sea trenches.

DC 10 – Pre-oceanic crustal destabilization adjacent to erupting "Fountains of the Great Deep" initiated a progressive "bottom-up" fracturing and volatile escape, causing collapse of the entire porous crust above.

DC 11 - Collapse of lineament-bounded pre-oceanic blocks undercut and crushed contiguous blocks, opening fracture conduits and releasing encapsulated volatiles to escape to the surface.

DC 12 – A thin layer of *basaltic magma* was squeezed laterally by compression of overlying collapsing oceanic crust in rolling pin fashion, destabilizing and initiating bottom-up collapse of adjacent lineament-bounded blocks.

DC 13 – Ocean basin collapse over-pressured *basaltic magma* below, forcing low-viscosity *basaltic* components to intrude upward into the collapsing oceanic crust above.

DC 14 – Porous crust collapse spread outward from deep sea trenches to collapse ocean basins worldwide during the Worldwide Flood Regression.

DC 15 – A substantial portion of rising liquid volatiles lost pressure and expanded into gases, which escaped into the atmosphere. Flood waters drained off submerged proto-continents into adjacent collapsing ocean basins.

DC 16 – Lunar tidal waves again beveled emerging continents as floodwaters receded and flowed rapidly across shallow surfaces, eroding the Great Unconformity across the top of unconsolidated Worldwide Flood deposits.

DC 17 – Flood regression and very early post-Flood erosion dislodged and transported a large volume of unconsolidated Flood materials off the continents, forming continental shelves.

DC 18 – Ocean basin collapse generated lateral expansion of the oceanic crust at depth, underthrusting adjacent continental rim areas. That overlap of continental and oceanic crust caused compressional uplift of continental rims.

DC 19 - Elevated continental rims inhibited seawater flow from continental interiors into collapsing ocean basins. Following the Worldwide Flood, shallow continental interior seas remained "perched" above ocean sea level, blanketing large mid-continent areas.

DC 20 – Continental crust was destabilized by tectonic and weathering processes extending from the surface downward during and following the Worldwide Flood. The pyroclastic rock matrix decomposed from exposure to groundwater and free oxygen. Encapsulated volatiles escaped from vugs, rose and comingled with the atmosphere and inland seas.

DC 21 – Weight-bearing capacity of the porous crust diminished as pyroclastic minerals decomposed and escaping volatiles ceased to facilitate overburden support. Lenses of weathered porous crust collapsed, extending compression fracturing of the porous crust below. Each collapse event exposed the next deeper layer of porous crust to matrix weathering decomposition, volatile leakage, and subsequent collapse.

DC 22 – Successively deeper layers of porous continental crust intermittently collapsed at a high frequency and intensity during the Post-Flood Millennium, depositing materials ranging from "Paleozoic" to "Pleistocene".

DC 23 – Intermittent collapse of continental basin crust released *pyromagma* volatiles and ash in shallow sea eruptions. Each *pyromagma* extrusion rose to the surface, spread outward, settled to the bottom, and consolidated into carbonate and shale formations. The local marine ecosystem was rapidly killed, buried, and fossilized by intermittent submarine eruptions of scalding pyroclastic volatiles and muds. Marine ecosystems quickly recolonized after each collapse event.

DC 24 – As the porous Antediluvian crust had solidified deeper, a progressional chemical sequence of different volatiles was encapsulated within the surrounding rock matrix. Intermittent top-down collapse of the porous continental crust released those volatiles in that same top-down order during the Post-Flood Millennium. Shale-rich muds progressively replaced carbonate-rich muds as porous crust collapsed at deeper levels.

DC 25 – That volatile release sequence produced a similar vertical order of marine lithologies and ecosystem remains in continental basins worldwide. However, the recolonization sequence progressed at different rates and times in separate continental basins. Marine ecosystems adjusted to the changing chemical components of inland seas.

DC 26 – Collapse under continental basin seas launched tsunami waves that periodically ripped out and deposited intertidal marine and lowland terrestrial ecosystem components on adjacent land. Terrestrial ecosystems ringing continental basins quickly recolonized after each tsunami event.

DC 27 – Continental basin seas and associated marine and terrestrial ecosystems periodically reconfigured, adjusting to the most recent subsurface collapse location. Sediment deposition progressively filled continental basins from basin flanks toward the basin center.

DC 28 – Continental roots continued to supply *hypomagma* beneath continental crust at a declining rate following the Worldwide Flood. As lateral hypomagma underplate flow reversed or rotated laterally from one direction to another, isostatic uplift tilted the crust above. Continental basin surfaces repeatedly tilted in changing directions, causing shallow inland seas to transgress and regress across hundreds of kilometers.

DC 29 – Isostatic tilting under continental interior seas breached topographic barriers, releasing catastrophic sheetflows far beyond continental basin boundaries.

DC 30 – "Paleozoic" carbonate deposition predominated during the early stages of continental basin collapse, while adjacent areas were relatively flat-lying and clastic import was minimal. However, as collapse compression proceeded to deeper levels, outward crustal displacement generated topographic uplift of surrounding areas. Accelerated erosion and sediment transport from those adjacent uplifting areas contributed an increasing proportion of clastic sediments.

DC 31 – Higher-elevation "Mesozoic" ecosystems routinely over-lapped lower-elevation "Paleozoic" and "Mesozoic" ecosystems as continental basins progressively infilled from continental rims toward continental basin centers. Clastic transport and deposition exceeded basin subsidence, causing continental basins to fill with sediments.

DC 32 – The final major continental basin collapse events generated the greatest compression-induced lateral displacement at the deepest depths. Mountain ranges "wrinkled" up along lines of crustal convergence between continental basins and where lateral continental basin collapse displacement butted into stabilized continental shields.

DC 33 – Rapid upheaval of continental interior mountain ranges accelerated erosion, transport, and deposition of clastic "Cenozoic" sediments outward across adjacent continental plains.

DC 34 – Rapid heat flow through the collapsed and intruded ocean floor sustained warm humid climates during most of the Post-Flood Millennium. Then, compression-induced mountain upheaval generated abundant volcanic activity. Volcanic ash collected in the upper atmosphere, reflecting sunlight back into space and reducing atmospheric temperatures worldwide. Water vapor "steamed" profusely from warm oceans into the cold atmosphere. Coastal areas became cloudy, dreary, and rain-drenched. Snow fell abundantly on cold continental interiors at high latitudes, accumulating as sunlight-reflecting glaciers to produce Ice Age climates.

DC 35 – Continental basin collapse, mountain upheaval, and volcanic activity subsided in frequency and intensity at the end of the Post-Flood Millennium. By that time, the porous crust had collapsed to its base in most continental basins. Atmospheric temperatures rose and oceans cooled. Evaporation of ocean water subsided. Cloud cover decreased and glaciers melted. Climates became increasingly sunny and arid. Modern ecosystems were established.

MAGMA SEGREGATION AND MIGRATION

Since creation, mineral density segregation has relentlessly progressed in the molten portions of Earth's interior. Low-density minerals rise. High-density minerals sink. The speed of position exchange depends upon magma viscosity and differences in density. *Pyromagma*, being of low density and low viscosity rose rapidly through higher-density magmas and arrived first in large quantities at the base of Earth's primordial crust. *Pyromagma* cooled and various minerals immediately began to solidify, forming a rapidly-deepening crystal melt mush. *Basaltic* magma, being of higher density and lower viscosity, rose more slowly and spread beneath the *pyromagma* layer.

Hypomagma is of slightly higher density than *pyromagma* and of lower density than *basaltic magma*. However, *hypomagma's* high viscosity slowed its rise toward the surface, and altered its flow dynamics. Viscous *hypomagma* segregated into larger and larger bodies deep in Earth's interior, until the *hypomagma* body was buoyant enough to rise. Smaller *hypomagma* bodies flowed into the wake of a larger rising *hypomagma* body and joined its "tail" to rise as a vertical cylinder-shaped continental root. Ascending *hypomagma* in continental roots arrived at the base of Earth's pre-continental crust in abundance during the Middle and Late Antediluvian Age.

During the Antediluvian Age, rising continental roots butted into the crustal ceiling, causing the Antediluvian crust above to bulge upward. Ascending *hypomagma* mushroomed outward, encasing low-density "bowl-shaped" *pyromagma* lenses and displacing *basaltic magma* laterally to destinations beneath adjacent pre-oceanic crust. *Hypomagma* was too viscous to spread uniformly beneath Earth's entire Antediluvian crust. Consequently, the *hypomagma* underplate only extended beneath approximately thirty percent of the Antediluvian crust to form pre-continental areas. *Hypomagma* contributions continued to underplate and tilt continental crust during the Post-Flood Millennium and tapered off thereafter.

A large, low-velocity continental root structure beneath the Parana province of Brazil is shaped like a vertical cylinder in the upper mantle. *"P- and S-wave velocity models exhibit remarkably similar features, a strong indication that the structures are authentic ... The low-velocity anomaly ... is roughly cylindrical in form, 300 km across and extending vertically from about 200 to at least 500-600 km depth (our resolution degrades severely at depths greater than 600 km) ... the plume ascends along a relatively narrow conduit through the lower-viscosity upper mantle and mushrooms beneath the Brazil/Africa continental lithosphere to a highly-flattened body perhaps 700 or 800 km in radius"* (VanDecar, et al., 1995, pp. 25, 30). (The authors are assuming Africa and South America split and drifted apart after the continental root formed.) **In contrast, Collapse Tectonics proposes proto-continents formed when low-density** *hypomagma* **rose up separate continental root conduits and mushroomed outward below the future South American and African continents.** (Continents were never "rafted" laterally for hundreds of kilometers across Earth's surface.)

MIDDLE ANTEDILUVIAN PRE-CONTINENTAL BASIN FORMATION

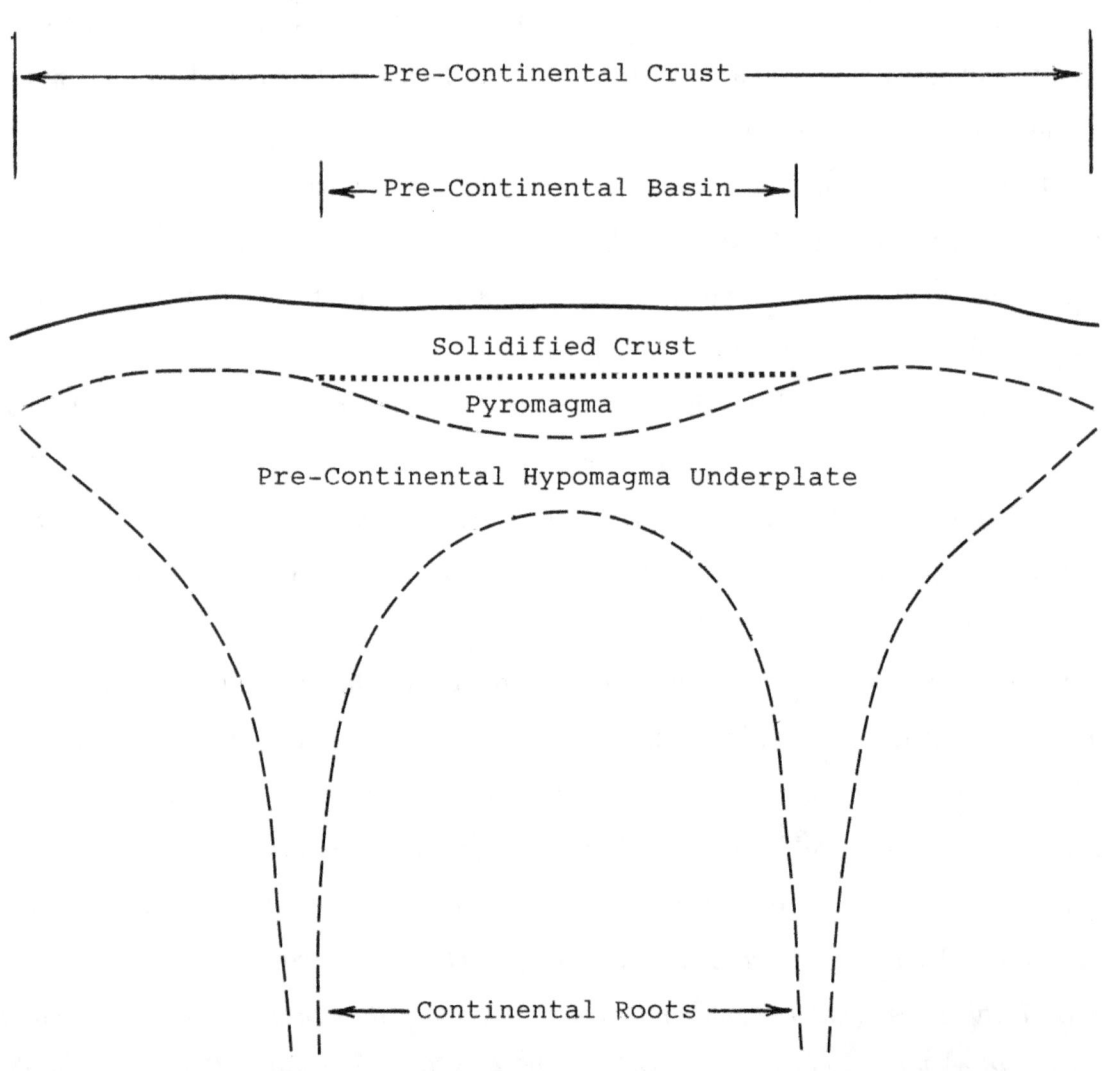

Pyroclastic Volcanics

In contrast to silicate-rich magmas, the comparatively higher melting point of some *pyromagma* mineral components produces the extrusion of *pyroclastic* ash instead of lava. To avoid subsurface solidification, a comparatively faster ascension speed is required through the upper crust. Because *pyromagma* has the lowest viscosity of the four magma categories previously described, *pyromagma* flows faster than other magmas. Also, *pyromagma* contains a greater proportion of liquefied gases (water, carbon dioxide, carbon monoxide, methane, ammonia, molecular hydrogen, and hydrogen sulfide). As *pyromagma* ascends and pressure decreases, those liquids flash into gas. Liquid-to-gas expansion accelerates *pyromagma* flow all the way up to the surface, causing a violent eruption.

Only five percent of dated eruptions have included a *pyroclastic* flow (Robinson, A., 1993, p. 95). However, *pyroclastic* eruptions are deadly. Mount Vesuvius has erupted more than 50 times during the past 2,000 years. Pompeii and Herculaneum were buried by mud, ashes, and air-borne rock fragments from one eruption. Pyroclastic ash enhances soil fertility, tempting people to reinhabit the area of recent annihilation (Robinson, A., 1993, p. 86).

The 1902 *pyroclastic* flow from Mount Pelee', on the Caribbean island of Martinique, silently annihilated about 30,000 residents of the Port of St. Pierre. Two months later, two British scientists were sailing past the ruins of St. Pierre. At dusk, a red glow from the Mount Pelee' summit became brighter and brighter. Then, an immense avalanche of dull red billowing gases swept down Pelee's flanks and across the ghostly ruins of St. Pierre. The boiling black cloud, coruscating with lightning, rushed out over the waters toward the sailboat. About a mile from the scientists, the cloud slowed and faded. A short distance away, the cloud raised and passed over the sailboat, dropping chestnut-sized stones, then pea-sized pellets, then ash. A faint smell of sulfuric acid lingered as the cloud broadened and covered the sky above (Robinson, A., 1993, p. 96).

Pyroclastic flows behave like fluids traveling on a cushion of air, sometimes leaving vegetation below unscathed. While watching Mount Pelee' from an observatory, the volcanologist, Frank Perret, described a pyroclastic flow as follows. *"The horizontal movement ... is due to an avalanche of a dense mass of hot, highly-charged and constantly gas-emitting fragmental lava ... extraordinarily mobile and practically frictionless because each particle is separated from its neighbor by a cushion of compressed gas. For this reason, its onward rush is almost noiseless"* (Robinson, A., 1993, p. 99).

Liquefied gases, including carbon dioxide and hydrogen sulfide, expand out of solution at a rate sufficient to support *pyroclastic* particles on an air-borne cushion of heavier-than-air gas. Theoretical analysis suggests that large *pyroclastic* flows may have moved as far as 60 kilometers from the vent at speeds as fast as 300 meters per second (Sparks, et al., 1978, pp. 1733-1736).

The 1980 eruption of Mt. St. Helens, Washington, U.S.A, included pyroclastic flows that are detailed as follows in United States Geological Survey Paper 1250. Many of the components of volcanic gases are unstable and rapidly form new compounds as they approach atmospheric temperatures and pressures (Casadevall et al., 1980, p. 190). The skin of *pyroclastic* pebbles is less vesicular (honeycombed) than the interior, indicating a volume expansion of about 15 percent after skin formation (Wilson and Head, 1981, p. 514). *Pyroclastic* pebbles rapidly decompose into soils upon exposure to surface conditions.

Due to its rapid flow, abundance of low-density components, and hotter primordial crust temperatures, *pyromagma* undoubtedly escaped to the surface in greater frequency and proportions during the Worldwide Flood and the Post-Flood Millennium. Since then, the proportion of higher-viscosity silicate magmas progressively increased to predominate in modern volcanic eruptions. **Due to selective prehistoric *pyromagma* depletion, Earth's early Antediluvian crust probably solidified primarily from *pyromagma* minerals.**

Pyromagma Vug Formation

Tiny liquid-filled cavities called "vugs" are encapsulated in modern igneous, metamorphic, and sedimentary rocks. Modern day vug formation usually amounts to less than five percent of rock volume, due to the ease of volatile escape to the surface through Earth's modern fracture system. Vug-filled rocks of deep mineral assemblages have been documented by Valley, et al, 1983; Glassley, 1983; Hansen, et al, 1984; Janaardhan, et al, 1982; and Selverstone, 1982. Fluid inclusions are predominantly carbon dioxide, with minor amounts of hydrogen sulfide, sulfur dioxide, carbon monoxide, and water. Modern ocean water contains an assortment of molucules similar to volatile fluid inclusions that were potentially sourced from within Earth's interior before escaping into seawater.

Slow upward flow of continental roots allows time for most *hypomagma* volatiles to escape to the surface before that *hypomagma* encounters lower temperatures and solidifies. However, if the *hypomagma* is injected rapidly upward and cools quickly, volatiles may be encapsulated in vugs within a quartz-rich rock matrix as a small percentage of rock volume.

How would that volatile encapsulation process function differently in a primordial setting? Rapid transfer of heat from the surface into the atmosphere (or seas) would have caused Earth's primordial crust to initially solidify downward very rapidly, forming an outer crustal shell. In modern times, most exsolutioned volatiles escape upward into the atmosphere through a highly-fractured crust. However, the modern tectonic processes that generate faulting and fracturing were not yet functioning. Earth's primordial crust should have behaved as a permeability barrier. Low-density volatiles rising from Earth's interior would have been blocked from escaping upward. Downward-solidifying magma would have quickly become oversaturated with volatiles. Supersaturated volatiles, being of lower density than the solidifying rock matrix could not be displaced downward. Therefore, volatile liquids would have been abundantly encapsulated in vugs within the downward-solidifying rock matrix, forming highly porous crust.

Earth's modern intensely faulted and fractured crust is attributed to the following tectonic processes. In pre-continental areas, the downward solidification of porous crust was interrupted by slowly-rising vertical shafts of viscous, low-density, quartz-rich *hypomagma* that would eventually solidify as "continental roots". However, "continental root" intrusives did not retain enough volatiles to form highly porous crust. That differentiated low-porosity *hypomagma*-sourded crust from porous *pyromagma*-sourced crust. Continental root underplating progressively thickened low-density materials under pre-continental areas, generating regional isostatic uplift. Resulting tensional faulting of adjacent pre-oceanic crust eventually penetrated the crust-forming magma zone to initiate the Worldwide Flood. Subsequent porous crust collapse sank ocean basins. Intermittent porous crust collapse went on to cause large-scale continental basin subsidence during the Post-Flood Millennnium. Compression at depth below collapsing continental basins generated outward displacement. Resulting lateral compression wrinkled up adjacent mountain ranges. Low-density volatiles rising from Earth's interior progressively depleted through time. Therefore, the supply of liquid volatiles available for encapsulation in vugs decreased as time passed. Those changing geomorphic conditions resulted in tectonic displacement that extensively fractured Earth's modern crust. **That highly-fractured crust fails to effectively impede the upward migration of liquid volatiles, thereby preventing the large-scale formation of new highly-porous crust.**

Crystal Melt Mush

Low-density, low-viscosity pyroclastic solids and liquids were first to rise from Earth's interior and occupy the molten magma zone underplating Earth's primordial crust. That crystal melt mush progressively solidified into porous crust as the *pyromagma* cooled. *Pyromagma* mineral components exhibit a variety of melting temperatures at specific pressures. Higher melting temperature components solidified first, followed by a succession of progressively lower-temperature components. Volatile liquids were continually exsolutioned as the *pyromagma* crystal-melt mush solidified. A pyroclastic rock matrix formed as volatiles were being expelled. The *pyromagma* below became over-saturated with volatiles. Exsolutioned volatiles were too low density to be displaced downward, so they accumulated in gaps between growing and overlapping crystals. Net porous crust density was similar to its *pyromagma* source below. Therefore, the *pyromagma* crystal-melt mush at the base of the crust was stable because the low-density volatiles were soon encapsulated as vugs near their exsolution point in newly-forming crust. A rock matrix (primarily apatite, calcite, gypsum, and halite) solidified around exsolutioned liquid volatiles (primarily water and carbon dioxide), capturing and preserving a sand grain-sized "honeycomb" of encapsulated vugs. Typical vug size is proposed to be smaller than three millimeters in diameter. Vugs are proposed to have occupied approximately one-third of total rock volume in the porous crust. **The weight-bearing capacity of the crust remained stable as long as the vugs remained sealed and pressures of volatile-filled vugs contributed to the weight-bearing capacity of the surrounding rock matrix.**

Antediluvian Crust Solidification

Let us presume the conditions generating the Worldwide Flood developed during the Antediluvian Age. In this case, low-density, low-viscosity pyromagma is proposed to rise to the base of Earth's crust, spread, cool, and solidify downward. Does the Biblical 1656-year Antediluvian Age allow sufficient time for the pyromagma crystal melt mush to cool and solidify into porous *pyroclastic* crust at depths down to the 25 kilometer depth proposed by Collapse Tectonics? Heat dissipation variables include magma convection, conduction of solids, and advection via groundwater circulation in lineaments. Any such estimation is necessarily oversimplified and speculative. However, heat dissipation estimates do provide a measure of confidence that the model parameters as proposed are reasonably possible. My thanks to Dr. Edward A. Boudreaux, (Professor Emeritus, University of New Orleans, Ph.D., MS, BS, Chemistry and Chemical Physics) for providing the following heat dissipation/porous crust solidification calculations.

BUDD'S PRIMORIDAL EARTH MODEL

by E. A. Boudreaux, PhD   (1/27/11)

Budd's model proposes that the primordial earth was such that the upper 1/3 of the crust would have had about a 10 km depth of vuggy composition containing bubbles of 1-3 mm diameter filled with H_2O, CO_2 and some dissolved minerals. Above this is a cap about 2 km depth composed of solid rock. The remaining 20 km or so is in a mushy state which is subjected to a pressure and temperature gradient.

In an effort to determine the depth of cooling over time, it is important to note that according to statistical mechanics, in the case of fluids, liquids and melts, the viscosity is the dominating factor in determining the thermal conductivity. A derived relation which provides thermal conductivity in its usual units (W/m K) is presented in equation (1).

$$\kappa = 10^{-3}\left[\frac{\eta \rho V C_p}{t^2 T^2}v\right]^{1/2} \quad (1)$$

K = thermal conductivity (J/s m K), η = viscosity (Pa -s = J s/m³), ρ = density (kg/m³), V = volume (m³), C_P = heat capacity (J/kg K), v = velocity (m/s), t = time (s) and T = temperature (K). Actually fluids and liquids the heat capacity should be at constant volume, C_V, but in these systems $C_V \approx C_P$ which is not the case with solids. However, the quantities ρ, V, v, t, T are not actual values, but only unit quantities for providing κ with the desired units.

Bubble Region

It will be presumed that the bubbles will behave as water in its super critical state, for which the average temperature is 638 K and average pressure is 235 bar (G. Ottonello, "Principles of Geochemistry", Columbia University Press, 1997, pg 487). It appears from data available that the viscosity increases in magnitude by about 2.5 x 10⁵ for every multiple unit increase in pressure. Thus from 1 bar to 235 bar pressure the viscosity should be about 235 x 2.5 x 10⁵ = 6 x 10⁷ Pa-s . The heat capacity at the given temperature is calculated to be about 332 (J/kg K) (G. Ottonello, pg 145). Using equation (1) and noting that everything under the square root cancels except η and C_P, the thermal conductivity would be:

κ = 10⁻³ [(6 x 10⁷) (332)(J²/s²m²K²)]^{1/2} = 141 (J/s m K).

The *thermal skin depth* provides the depth of cooling as a function of the rate of heat transfer from the medium. This is given by:

$$d = \left[\frac{\kappa}{\rho C_P}(t)\right]^{1/2} \quad (2)$$

The density of supercritical water at its critical point is 333 kg/m³ (G. Ottonello, pg 487), thus substitution of appropriate data into equation (2) yields the following:

$$d = \left[\frac{141}{333 \times 332}(m^2/s)xt\right]^{1/2}$$

And upon substituting the proposed 1656 years of cooling in Budd's model, the result is:
d = [1.3 x 10⁻³ (m²/s) (1650 yr) (3.16 x 10⁷ s/yr)] = **8.2x10³m** or 8.2 km

Mushy Mineral Matrix Region

The model maintains that the 20 km region below the bubbles is a mushy mix of non-silicate minerals. The predominant ones would be apatite, calcite, gypsum, and halite. It is difficult to obtain any reliable information on the relative abundances of these minerals. However we can glean a general trend by obtaining relative ratios of the pertinent elemental abundances, together with the relative solubilities. The trend should be that those having higher elemental abundance ratios and lower solubilities, will be the more abundant minerals. The oxygen is considered a constant since its abundance is so much greater than any of the other involved elements. Table 1 contains the required data.

Table 1

Mineral	Formula	Elemental Ratios	Normalized to Unity	Solubilities(g/100g H_2O)	% Abundance[a]
Apatite	$Ca_5(PO_4)_3F$[b]	3P/5Ca = 3(0.3)/3(4) = 0.05	0.05/0.068 = 0.74	10^{-4}	74
Calcite	$CaCO_3$	C/Ca = 0.018/4 = 0.005	0.005/0.068 = 0.07	6×10^{-3}	15
Gypsum	$CaSO_4(2H_2O)$	S/Ca = 0.03/4 = 0.008	0.008/0.068 = 0.12	0.2	9
Halite	NaCl	Cl/Na = 0.013/2.8 = 0.005	0.005/0.068 = 0.07	28	2[c]

a. See text for the source of these values
b. F is not considered significant for these purposes
c. Other minerals in lesser amounts may be considered as a part of this figure.

It is commonly accepted that the earth's upper crust is 75% sedimentary. It is further stated that up to 15% of this crust contains calcite (www.physicalgeography.net/fundamentals/10da.html). Hence, the normalized calcite ratio, 0.07 is equivalent to 15% abundance. Since gypsum is 33 times more soluble than calcite (see Table 1) then the relative ratio should be 0.12 x0.33 =0.04, Since a ratio of 0,07 for calcite is equivalent to 15% abundance, then a ratio of 0.04 for gypsum is equivalent to 9% abundance. The remaining 2% is assigned to halite plus lesser amounts of other minerals.

The weighted averages of density and heat capacity are contained in Table 2.

Table 2

Mineral	Density (kg/m³)	Wtd. Ave.	Heat Capacity(J/kg K)*	Wtd. Ave.
Apatite	3290	3290 x 0.74 = 2435	~ 840	840 x 0.74 = 622
Calcite	2710	2710 x 0.15 = 407	~ 970	970 x 0.15 = 146
Gypsum	2329	2320 x 0.09 = 209	~ 669	669 x 0.09 = 60
Halite	2170	2170 x 0.02 = 43	~ 906	906 x 0,02 = 18
		Total = 3094		Total = 846

*Temperature ~ 499 C⁰

So in calculating the thermal skin depth for this system a net density of 3094 kg/m³ and a net heat capacity of 846 j/kg K will be employed. It is difficult to know what should be the magnitude of the viscosity for this mushy system under high pressure. Most experimental data on melts involves silicates and show that viscosities ranging from 10^4 to 10^{14} are possible, depending on pressure and temperature. If the geometric average of this range is taken then 10^9 is the magnitude. But here again this seems rather low, since the viscosity of super critical water has a magnitude of 10^7 at a pressure 3 x 10^2 bar. However, at the lower depths in the crust the pressure increases by another order of magnitude, and it is likely that the magnitude of the viscosity would increase by more than 10^2 (I,e. from 10^7 to 10^9). Consequently, the geometric average between 10^9 and 10^{14} which is 3×10^{11}, should be a more reasonable magnitude for the viscosity. Substituting into equation (1) we find:

$$\kappa = 10^{-3} [(3 \times 10^{11} \text{ Pa-s}) (840 \text{ J/kg K})]^{1/2} = 1.6 \times 10^4 \text{ (J/s m K)}.$$

Further substitution into equation (2) yields:

$$d = \left[\frac{1.6 \times 10^4}{(3094)(840)} (m^2/s) \times 5.2 \times 10^{10} s \right]^{1/2} = 1.8 \times 10^3 m = 18 km$$

Combining the initial result for the bubble region with this result for the mushy region: **8 km + 18 km = 26 km.** This is a reasonable result in light of all the uncertainties involved. **Thus it appears that Budd's proposed model is vindicated.**

Pre-Continental / Pre-Oceanic Crust Comparison

Pre-oceanic crust progressively solidified downward with negligible magmatic contributions from Earth's interior. In contrast, pre-continental crust was underplated by low-density *hypomagma* rising from Earth's interior through "continental roots". Upon bumping into the solid crust above, *hypomagma* branched outward laterally. Deep *pyromagma* pockets remained adjacent to the *hypomagma* branches. Cooler temperatures at the interface between solid crust above and *hypomagma* flow below caused *hypomagma* to become progressively more viscous and begin to solidify. Adjacent areas occupied by *pyromagma* were also solidifying deeper and deeper. The low-density continental root column caused the crust above to isostatically rise. Consequently, extension of laterally flowing *hypomagma* branches was slightly down-dip. That inhibited outward extension of the *hypomagma* branches across long distances. Continuation of the *hypomagma* supplementation process throughout the Antediluvian Age generated thickened areas of low-density crust that would later become continents.

Downward-solidifying crust was isostatically-balanced and stable during the Early to Middle Antediluvian Age. However, by the end of the Antediluvian Age, continental roots had supplemented adjacent areas with enough low-density *hypomagma* to form a thick, underplate beneath pre-continental crust. Precontinental crust was isostatically buoyed upward relative to preoceanic crust.

Rising *hypomagma* from Earth's hot interior sustained a higher heat flow near upwelling continental root diapirs. Therefore, pre-continental crust cooled and solidified at a slower pace than pre-oceanic crust. That resulted *pyromagma*-sourced crust being somewhat thinner in pre-contingneal areas than in pre-oceanic areas. By the end of the Antediluvian Age, porous crust extended to depths up to 20 kilometers in pre-continental areas and averaged 25 kilometers in pre-oceanic areas. The pre-continental *hypomagma* underplate ranged from 3 to 18 kilometers in thickness, depending upon time of first arrival of *hypomagma* sourced by each continental root and the lateral migration paths formed around remnant *pyromagma* bodies. **In conclusion, pre-continental crust was underplated by a progressive accumulation of low-density *hypomagma* sourced by upwelling continental roots.**

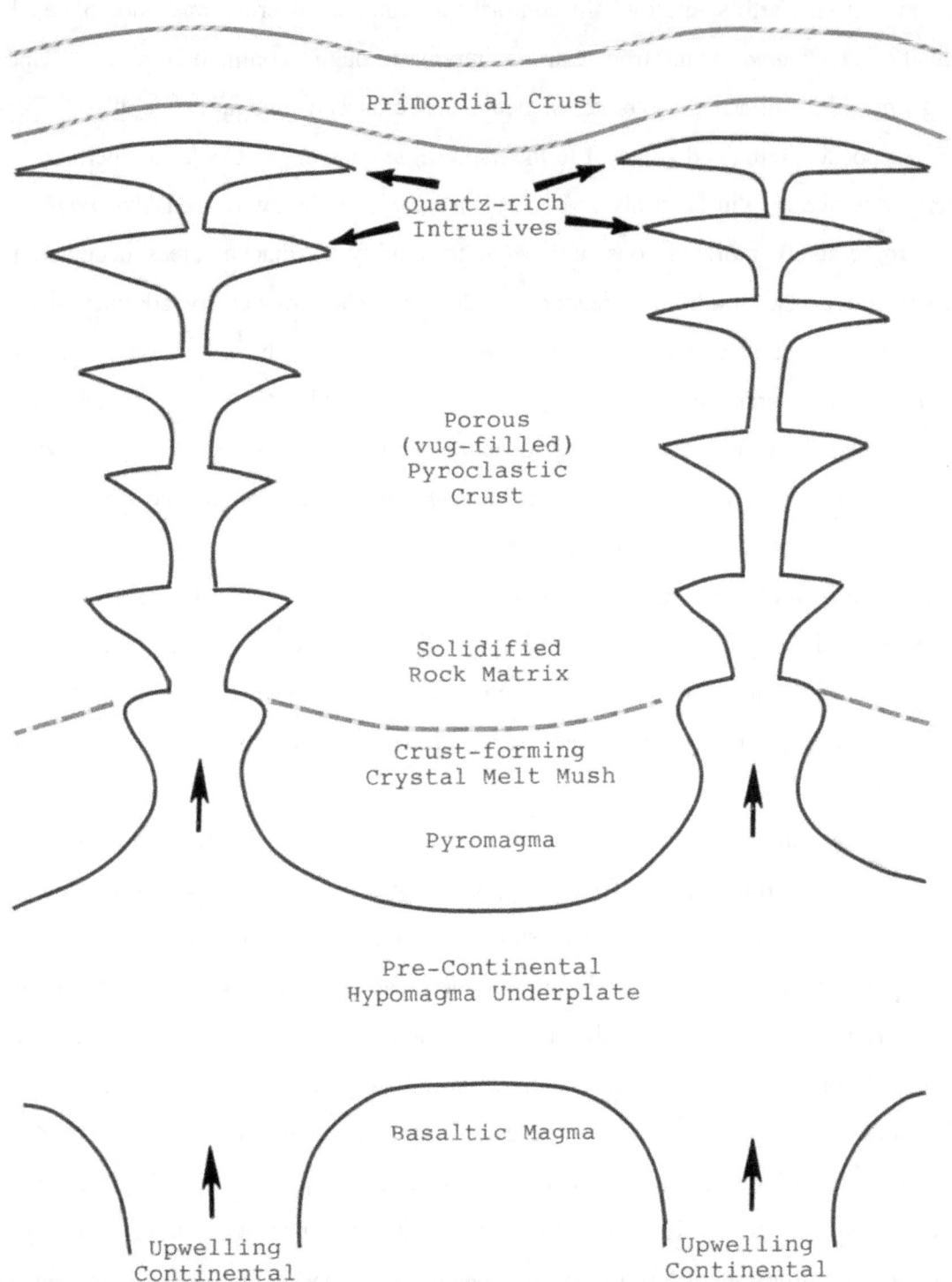

During the Worldwide Flood, the pre-continental crust was less vulnerable to tectonic disruptions than the pre-oceanic crust, because the viscous, low-density *hypomagma* underplate acted like a "silica-gel flotation cushion", stabilizing the pre-continental crust above. However, oceanic crust underthrusting fractured and destabilized some continental rim areas. Worldwide Flood erosion removed impermeable primordial crust in other continental areas, exposing underlying porous crust to mineral decomposition leading to structural fatigue.

Pre-Oceanic Crust Destabilization

The fragile Antediluvian crust was stable due to the separation between the base of the downward-solidifying porous crust and lineament fractures above. That separation constituted a downward-shifting permeability barrier capable of sustaining differential pressures above and below in a stable relationship, providing no tectonic disruption breached that barrier. However, the Antediluvian crust was soon to be irreversibly destabilized at the onset of the Worldwide Flood.

Basaltic and *ultramafic magmas* beneath pre-oceanic porous Antediluvian crust were of higher density than the adjacent *hypomagma* underplate accumulating beneath pre-continental areas. Therefore, pre-continental crust was buoyed upward by isostatic displacement. That uplift stretched the adjacent pre-oceanic crust parallel to the pre-continental crust boundary. The greatest tension occurred parallel to the Pacific Rim. **Tensional faults extended downward, approaching dangerously close to the volatile-rich, crust-forming magma zone below.** Meanwhile at an unknown location on Earth's surface, God was closing the door to Noah's Ark (Genesis 7:16).

Triggering the Worldwide Flood

At the end of the Antediluvian Age, isostasy-induced tensional faults paralleling pre-continental boundaries extended downward to penetrate the base of the crust in what would become the Pacific region. Overpressured volatiles within the crust-forming magma zone flowed laterally beneath the pre-oceanic crust toward open faults, raced upward, and jetted water, carbon dioxide, pyroclastic gases, and volcanic ash several kilometers into the atmosphere. Volcanic ash provided abundant particle surfaces for water vapor to condense upon, *"and rain fell on the earth forty days and forty nights"* (Genesis 7:12, NIV).

The pre-oceanic crust sank as trapped volatiles escaped from the crust-forming magma zone. Surface waters flowed toward and inundated subsiding fountains within weeks. "Fountains of the Great Deep" became submerged. Surface waters intercepted direct atmospheric contributions of fountain water and volcanic ash. Torrential rains ceased when volcanic ash was purged from the atmosphere by rains and no longer provided condensation surfaces for raindrops to form.

Hypomagma underplating the pre-continental crust was too viscous to flow beyond pre-continental boundaries, and did not contribute to the "Fountains of the Great Deep".

Deep Ocean Trenches

Fissures in Earth's Antediluvian crust tore open across thousands of kilometers paralleling pre-continental boundaries at the onset of the Worldwide Flood. *Pyromagma* volatiles raced up open fissures and accelerated to phenomenal velocities, as pressure decreased and rising liquids expanded into gas. **Fissure walls were scoured back, carving deep ocean trenches.**

In areas where pre-continental boundaries abruptly change direction, trenches commonly extend in linear fashion outward into oceanic crust. However, the isostatic displacement stress that caused the trench to split open, remained parallel to the pre-continental crust boundary. Therefore, any trench extending outward from a continental boundary curved back toward the pre-continental boundary alignment. Greater curvature of an oceanic trench extension, indicates greater tension (pulling apart) of the crust between the continental boundary and the oceanic trench. Extremely high tension associated with greater curvature, tore open the Mariana Trench to the greatest depth on Earth's surface.

Trench formation continued at a decreasing pace after the initial 40 days of the Worldwide Flood. However, volatile depletion of the crust-forming magma zone slowed flow rates and corresponding erosion. Lower pressures associated with the modern fractured crust prevented subsequent cataclysmic eruptions at a magnitude resembling the "Fountains of the Great Deep".

OCEAN TRENCH FORMATION

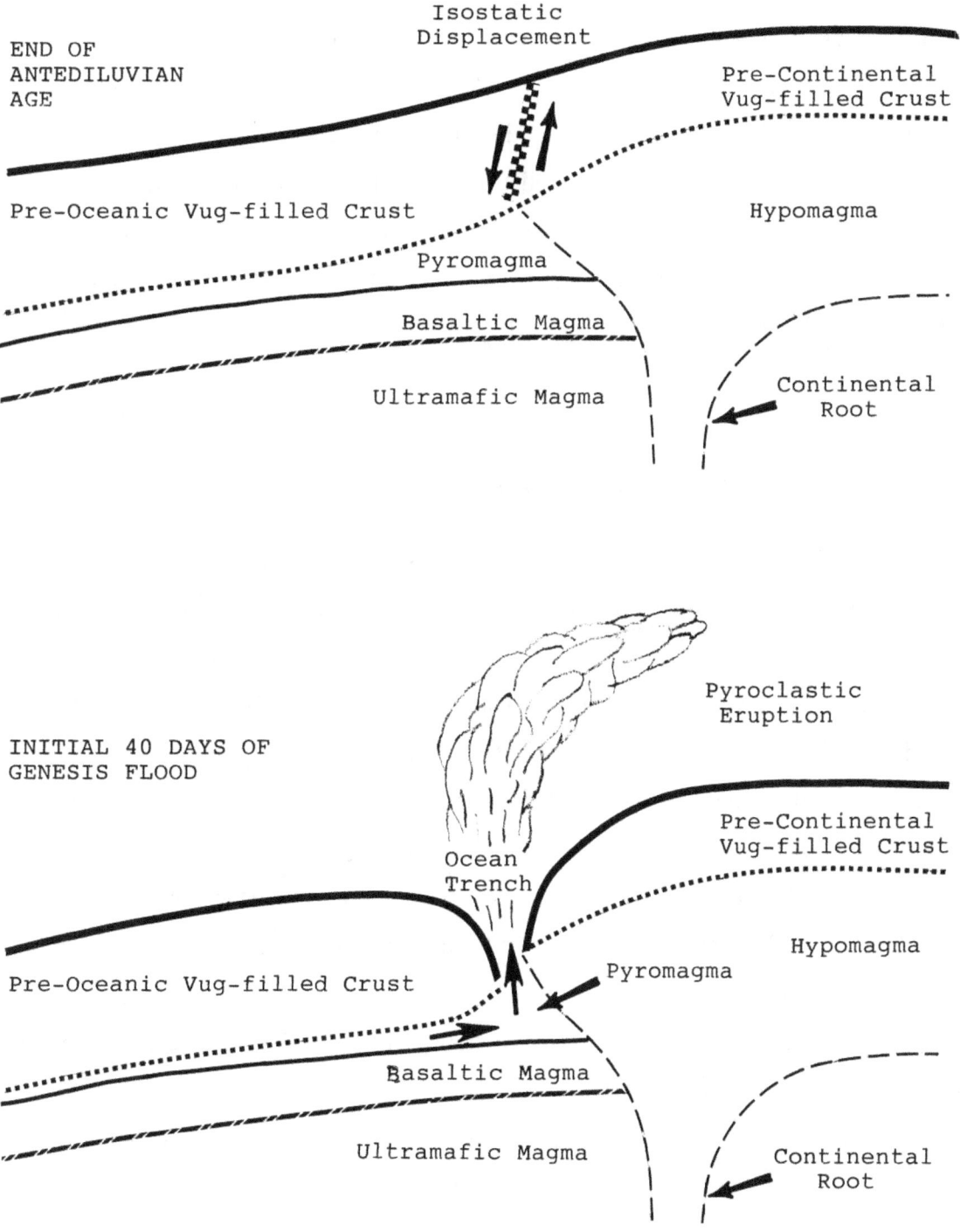

The Zodiac Fan was deposited by rapid sediment transport and deposition during the Inundation Stage of the Worldwide Flood. The Aleutian Trench bisects the Zodiac Terrigeneous Fan. Therefore, the Zodiac Fan was deposited prior to the opening of the Aleutian Trench. The Aleutian Trench is interpreted as a Regression Stage and/or early post-Flood feature.

Ocean floor rifting continues at a much slower pace in modern times. The continuing degasification of Earth's shrinking interior continuously over-sizes Earth's crust, producing subtle lateral compression undulations. *Basaltic magma* flows beneath the crust upslope toward the crest of those undulations, displacing higher-density *ultramafic magma*. Thickening *basaltic magma* isostatically uplifts and stretches Earth's crust, opening tensional rifts along the crest of flexures. A 500 kilometer section of the Woodlark Basin Rift, on the continental shelf northeast of Australia, appears to have recently burst open (Taylor, B., et al., 1995; Goddard, 1995; Austin, et al., 1994).

Proterozoic Sedimentary Deposition

Early and Middle Proterozoic sedimentary formations are a continuation of Worldwide Flood Inundation Stage deposition that was less vulnerable to igneous melting and metamorphic alteration than deeper "Archean" strata. **Catastrophic Flood tectonics eroded pre-continental topographic highs and transported those materials along with freshly-extruded volcanic materials to be deposited in adjacent topographic lows.** Proterozoic sedimentary formations are typified by quartzite-carbonate-shale assemblages with scattered dikes, sills, and volcanic extrusives. These unconsolidated Inundation Stage sediments were sometimes tilted, elevated, and sheared off during the Regression Stage of the Worldwide Flood. Those eroded materials experienced multiple episodes of erosion, transport, mixing with other materials, and deposition in post-Flood sedimentary formations. A substantial portion of eroded Proterozoic materials were transported to adjacent oceans, where they settled out as continental shelf deposits.

Pyroclastic ash contributions sourced by submerged fountains were captured by sea water instead of being expelled directly into the atmosphere. The pyroclastic ash component sank and slowly settled onto the sea floor, comingling with eroded and laterally transported

Flood deposits. Rain condensed upon and scrubbed pyroclastic ash from the atmosphere. Basalt-intruded crust and hot ocean waters progressively cooled during the Post-Flood Millennium.

Porous Crust Collapse Potential

Pyromagma solidification formed porous crust averaging two-thirds apatite/dolomite/calcite/gypsum rock matrix and one-third liquid-filled vugs by volume. Worldwide Flood Regression Stage collapse expelled encapslulated liquids, producing an average 30 percent decrease in pre-oceanic crust volume. **Earth's crust above collapsing zones correspondingly sank to form ocean basins.**

By the end of the Antediluvian Age, pre-oceanic porous crust extended to a depth averaging 25 kilometers (km). Subsequent oceanic crust collapse averaged 7.5 km. Subsidence was greatest near the point of collapse origin. However, as ocean basin collapse spread, horizontal compression caused Earth's crust to expand laterally. The increase in horizontal compression resistance within Earth's crust prevented the final Arctic Ocean basin collapse area from sinking as deeply as the oceans basins that collapsed earlier.

Pyromagma-sourced pre-continental crust extended to a maximum depth of approximately 20 km beneath future continental basins. Continental basins were circumvented by laterally-extending branches of *hypomagma.* Continental basin collapse compression is approximately 18 km x 0.3 = 5.4 km. Thicker stratigraphic sections are attributed to Worldwide Flood tectonics, eroding topographic uplifts and infilling tectonic downwarps. The Post-Flood Millennium was characterized by frequent and intense subsurface collapse of porous crust, causing continental basins to subside.

The cause-and-effect mechanism overpressuring and transporting intermittent pulses of hydrothermal fluids correspond to the deeper stages of continental basin porous crust collapse during the latter centuries of the Post-Flood Millennium. Collapse-induced overpressuring of the crust forming magma zone below could not be displaced downward into underlying denser magmas. Lateral compression redirected crustal displacement outward and upward from beneath continental basins, generating fractures and wrinkling up mountain ranges in adjacent areas. Low-density, low-viscosity hydrothermal fluids raced through freshly-opening fractures until their upward flow path dispersed into permeable unconsolidated sediments at a few kilometers of depth.

The Worldwide Flood held clays in suspension longer than sands. Therefore, a thick sequence of permeable sands were consistently deposited first above bedrock and sequentially capped by silts and clays (normal graded bedding). Rising overpressured hydrothermal fluids typically penetrated and fanned-out in those permeable water-filled sands. However, clay-rich layers above acted as a permeability barrier, diverting those hydrothermal fluids laterally outward through the underlying sand layer. Therefore, widespread layers of water-filled sands provided the primary lateral distribution conduits for intruding hydrothermal fluids.

That reasonably explains the so-called "space problem" regarding the tabular three-dimensional geometry of granite plutons. *"The majority of plutons so far investigated using detailed geophysical (gravity, magnetic susceptibility, and seismic) survey appear to be flat-lying sheets to open funnel-shaped structures with central or marginal feeder zones"* (Snelling, 2009, Vol. 2, pp. 989-990; also see Evans, D.J. et al, 1994; Ameglio, J.-L et al, 1997, pp.199-214; Amegilo, L. and J.-L Vigneresse, 1999, pp. 39-54; Petford, N. and J.D. Clemens, 2000, pp. 180-184).

Hydrothermal fluids rising through open fracture conduits below flared out upward and intruded adjacent water-filled unconsolidated sand layers. Higher-viscosity hydrothermal components began to cool, congeal and accumulate directly above the fracture feeder zones. The progressive accumulation of lower-viscosity hydrothermal components reduced permeability and diverted intruding fluids outward within adjacent sand layers above fracture feeder zones. The zone above the fracture feeder was subject to the hottest temperatures and the most intense melting and comingling of unconsolidated sands and low-viscosity hydrothermal components. Melting and mineral dissolution of bedrock surrounding the feeder fracture extended downward below the sand layer in a conical shape. Subsequent solidification of residual higher-viscosity materials formed the characteristic upward-widening "funnel shape" of very hard granite.

Lower-density hydrothermal fluids bypassing the accumulating "funnel" flowed outward from the from the fracture feeder through the unconsolidated sand layer. Overpressured hydrothermal fluid flow destabilized the sand grain packing arrangement, erasing sedimentary deposition structure. However, the lateral transport of those intruded sediments was negligible. Connate waters in the intruded sediments boiled, agitating the hydrothermal-fluid/melted-sediment slurry until excess connate water was driven off. Hot hydrothermal fluids melted sediment grain surfaces and more. Heat dissipation associated with boiling and evacuation of excess connate waters caused the sediment slurry to congeal. Sediment grains and residual hydrothermal minerals cooled and crystalized in sequence according to their respective solidification (a.k.a. melting point) temperatures, forming the characteristic granular granitic texture.

High-grade metamorphism relates to distance from the fracture feeder and unconsolidated sediment permeability. As distance from the fracture feeder increased, hydrothermal fluids progressively cooled by contact with unconsolidated sediments along that extending flow path. Agitation from boiling of connate water subsided, suppressing intergranular movement and preserving some of the original sedimentary mineral fabric before solidifying. However, overpressured hydrothermal flow mobilized, warped, and folded sedimentary materials prior to solidification into high-grade metamorphic rocks.

Lower-permeability clay beds and lenses were penetrated more slowly by lower-viscosity hydrothermal components. Index minerals (chlorite, biotite, garnet, staurolite, sillimanite, K-feldspar) are primarily diagnostic of changing conditions as hydrothermal fluids expelled from overpressured sands and solidifying granites intruded adjacent clay-rich sediments. Lensing mixtures of variably-permeable sands, silts, and clays were metamorphically altered and solidified into gneiss.

Post-solidification cooling caused rock shrinkage and tensional fracturing. Subsequent tectonic uplift events tilted and more extensively fractured the igneous and metamorphic layers above. Fracturing tended to circumvent the thicker, harder funnel-shaped granitic zones above fracture feeders.

Lower-viscosity hydrothermal fluids within the crust-forming magma zone progressively depleted as time passed. Consequently, expelled hydrothermal fluids became more viscous. By the end of the Post-Flood Millennium, unconsolidated sediments near preexisting fracture feeders had already been intruded and consolidated into granite, schist, and gneiss.

Fracture feeders were re-opened with each new tectonic uplift pulse. Overpressured, late-stage hydrothermal fluids were usually transported along those same fracture conduits and injected into bedrock fractures near the original fracture feeder sites. Therefore, granites are highly-concentrated around fracture feeder sites.

"Granites more frequently than any other rock have pegmaties genetically associated with them" (Dana, E.S. and J.D. Dana, 1959, p. 522). The residual low solidification temperature, quartz-rich liquid expelled from solidifying batholiths penetrated and widened fractures above before solidifying into pegmatite veins and dikes. Thick pegmatite dikes are typically characterized by the growth of larger crystals of quartz, feldspar and mica than the surrounding materials. The longer time of heat retention extended mineral segregation and crystal growth time to form larger crystals. In contrast, thin pegmatite veins lost heat more rapidly, resulting in smaller crystal sizes.

Subsurface conditions progressively changed as Earth aged. Deep porous crust collapse events in continental basins 1.) compressed and selectively squeezed out low-density, low-viscosity pyroclastic-rich hydrothermal fluids from the crust-forming magma zone below, 2.) transported those overpressured fluids laterally through outward displacement-induced fractures, 3.) intruded overpressured hydrothermal fluids into unconsolidated Worldwide Flood sediments at few kilometers of depth, and 4.) metamorphically altered those unconsolidated sediments into granite, gneiss, and schist.

Optimum granite, gneiss, and shist-forming conditions occurred during the closing centruies of the Post-Flood Millennium in association with mountain-building events. Thereafter, pyroclastic volatiles were too depleted in crust-forming magma to generate highly-porous crust. Lenses of remnant porous crust continue to collapse in modern times. However, granites are not observed to be forming today because the set of variables facilitating granitic rock formation are no longer in play. **In summary, most so-called "Precambrian" granites, gneisses, and schists solidified from unconsolidated Worldwide Flood sediments that were hydrothermally altered in place at a few kilometers of depth during the Post-Flood Millennium.**

OCEAN BASIN COLLAPSE

"The waters stood above the mountains. But at your rebuke the waters fled,
... they flowed over the mountains, they went down into the valleys"
(Psalms 104: 6b, 7a, 8a, NIV)

Antediluvian crust is herein proposed to have solidified from *pyromagma* to an average depth of 25 km in pre-oceanic areas. Approximately two-thirds of rock volume was primarily composed of an apatite/dolomite/calcite/gypsum rock matrix. The remaining one-third of rock volume consisted of liquid-filled vugs. Average oceanic crust collapse subsidence is 25 km – 2 km (primordial crust) = 23 km porous crust x 0.3 (volatile space compression) = 6.9 km. Ocean basin collapse is potentially accounted for because average modern ocean depth is only 3.79 km. Modern ocean waters were predominantly sourced by the crust-forming crystal melt mush zone via erupting "Fountains of the Great Deep" and by Regression Stage volatile contributions escaping from collapsing oceanic crust. Antediluvian lakes provided a minor sea water contribution. Abundant Worldwide Flood release of subterranean carbon dioxide and phosphorous minerals facilitated prolific post-Flood plant growth.

Collapse of porous pre-oceanic crust was instigated and sustained by the following processes:

1.) Isostatic tension induced pre-oceanic crust fracturing parallel to pre-continental boundaries.

2.) Pre-oceanic crust fracturing, as the crust sank into the area vacated by *pyromagma* volatiles escaping via the "Fountains of the Great Deep".

3.) Melting of the base of the pre-oceanic crust by contact with the hotter *basaltic magma* layer below.

4.) Release of liquid volatiles from vugs reduced the weight-bearing capacity of the pre-oceanic porous crust matrix, thereby facilitating collapse.

5.) Volatile absorption by *basaltic magma* initiated melting and lowered viscosity, thereby enhancing intrusive mobility.

6.) Horizontal compression displacement of collapsing lineament-bounded blocks undercut and crushed the lower portion of adjacent porous crust.

7.) Collapse compression squeezed deeper *basaltic magma* horizontally as an advancing subsurface magma swell, destabilizing adjacent uncollapsed crust.

8.) *Basaltic magma* swell uplift fractured porous crust above, sustaining lateral collapse advance throughout the Regression Stage of the Worldwide Flood.

OCEAN BASIN COLLAPSE

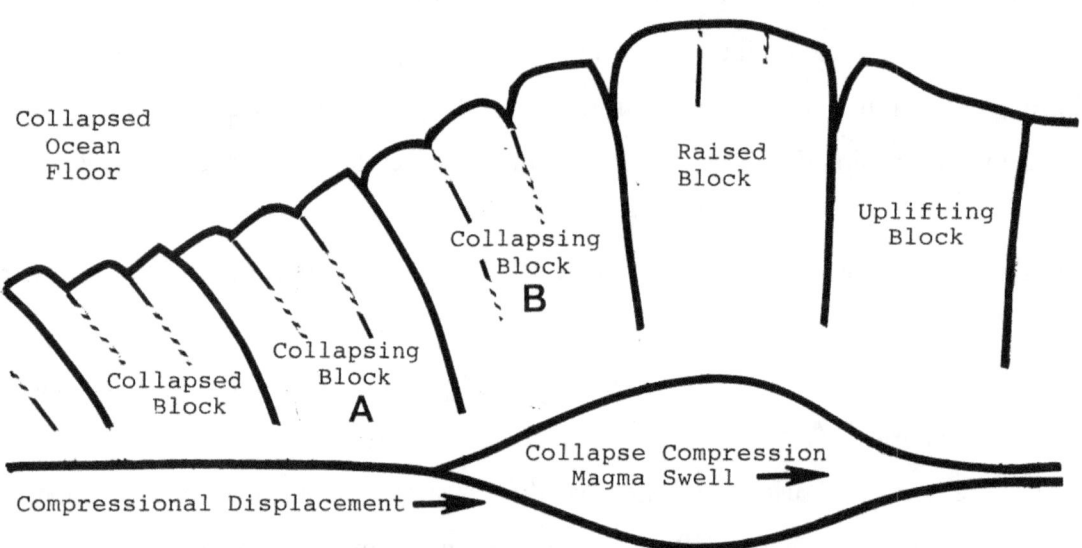

Vertically Exaggerated Cross Section

Collapse of the Antediluvian crust began on the oceanic side of deep sea trenches, where tensional destabilization of pre-oceanic crust was most intense. Porous crust collapse was initiated at the base of the crust, which progressively destabilized the crust upward toward the surface. Base-to-surface collapse of each lineament-bounded block of porous crust extended outward and upward from the base of each block.

Collapsing crust at one location undercut and destabilized the adjacent lineament-bounded block, causing collapse to continuously spread laterally from block to block, forming the deep ocean basins. Ocean basin collapse began in the North Pacific, spread into the South Pacific, advanced westward through the Indian Ocean area, turned north into the South Atlantic, spread into the North Atlantic, and extended into the Arctic Ocean. The topographic configuration of post-Flood ocean basins closely resembled modern oceans.

Initially, collapse of oceanic crust was tension-dominated. That tensional, early ocean basin collapse was characterized by deep sea trenches and relatively flat seafloors, interrupted by large volcanic cones. As ocean basin collapse spread around Earth's surface, downward collapse compression was redirected horizontally. Ocean basin collapse altered to produce compression-dominated orthogonal seafloor topography, with fewer and smaller volcanic cones.

Orthogonal Seafloor Topography

Collapsing oceanic crust produced compression flattening and outward displacement encroaching upon the base of adjacent lineament-bounded blocks. As seafloor collapse spread, horizontal compression progressively increased in the same direction that ocean basin collapse advanced. Lateral compression displacement caused entire lineament-bounded oceanic blocks to collapse one after another.

Collapse compression flattening and resultant horizontal displacement was greatest at the base of each block. Block "A" collapsed and undercut Block "B", causing the top of Block "B" to tilt back and "lean on" the sinking Block "A". That facilitated crushing of the upper portion of Block "A". As collapse advanced from block to block, each successively-collapsing block was tilted back onto the block behind, resembling a row of fallen dominoes.

In this case, however, the lower portion of the crust was displaced further forward than the upper portion of crust.

Orthogonal oceanic crust is characterized by *"multiple, closely-spaced fault scarps that strike parallel to each other and are easily detectable in long-range sidescan sonographs"* (Wezel, 1992, p. 429). **Ridges and valleys formed as collapsing lineament-bounded blocks tilted back, away from the direction of collapse advance.** Those tilted blocks aligned perpendicular to the direction of collapse advance.

"Crustal subsidence in turn appears to induce earthquakes, fracture zones, mid-ocean ridges, and many of the patterns of sediments and anomalies..." (Dillon, 1974, p. 217). Mid-ocean ridges extend along the path of advancing ocean basin collapse. The approximate fit between the continents across the Atlantic Ocean is explained by crustal subsidence framed by orthogonal fractures, without requiring seafloor spreading (Anfiloff, 1992, pp. 75-109). The sequential progression of ocean basin collapse generated fracture zones oriented perpendicular to Mid-oceanic ridges.

Paleomagnetic Anomalies

Alternating stripes of seafloor magnetism are mapped as subtle variations in magnetic intensity - not positive-negative magnetic reversals (Chatterjee and Hotton, eds., 1992, p. 171). Localized polarity reversals weaken magnetic intensity strength. Plate Tectonics assumes these subtle magnetic intensity variations correspond to prehistoric reversals in Earth's magnetic field as new oceanic crust was forming outward from Mid-ocean ridges through time. Slightly weaker magnetic intensity stripes are presumed to represent seafloor crust solidification when Earth's magnetic field was opposite to the modern North/South orientation.

In contrast, Collapse Tectonics advocates the simpler concept that these subtle magnetic intensity variations are fracture-controlled. Magnetic anomaly stripes intersect continents in at least 15 places and dive below Precambrian or Phanerozoic rocks. Magnetic anomaly stripes are concentric around Precambrian continental shield areas. That implies the rejuvenation of the Precambrian fracture system is directly related to these magnetic anomaly stripes (Meyerhoff and Meyerhoff, 1974).

"Inspection of the map of magnetic intensity shows that the magnetic anomaly pattern in the (Mid-Atlantic Ridge) *area ... may be related mainly to zones of fresh basalts, alternating with hydrothermally altered zones which perhaps surround fracture zones"* (Luyendyk and Melson, 1967, pp. 148-149). Therefore, alternating stripes of slightly weaker and stronger magnetic intensity are more simply explained by secondary variations in iron-rich mineral injection and higher heat flow upward through seafloor fractures.

Island Arcs

Trenches and volcanic island arcs reveal an interdependent relationship in position. The Aleutian, Kuril, and Mariana Trenches are tensionally weakened areas of oceanic crust that partially encircle volcanic island arcs. The deepest Mariana Trench features the tightest encircling arc with the Mariana Island chain being most distant from the Trench.

Why does the concave side of ocean trenches face adjacent continental crust? Increasing horizontal tension between pre-continental crust and pre-oceanic crust caused pre-oceanic crust to rift apart. Tearing apart of the pre-oceanic crust began at the point of greatest tension and extended somewhat parallel to adjacent pre-continental/pre-oceanic boundary. However, the trench-forming rift extended outward in both directions along an arc-shaped path, veering back toward the adjacent pre-continental boundary. Greater tension produced greater arc curvature as exemplified by the deepest rift in the oceanic crust at the Mariana Trench.

What geomorphic process caused volcanic arcs to be offset to the concave side of the adjacent trench? Tensional fracturing was greatest along the concave side of the trench, generating more efficient subcrustal volatile flow from that partially-encircled area into the trench and upward to the surface to be expelled as the "Fountains of the Great Deep". That caused the adjacent crust to sag abruptly along the concave side of the trench. That sag was greatest at the trench, generating a uniform upward tilt of the crust away from the trench on the concave side. Where tilted crust met horizontal crust, a subordinate tensional fracture set extended from the surface downward. That island-forming fracture set ran parallel to the concave side of the arc-shaped trench.

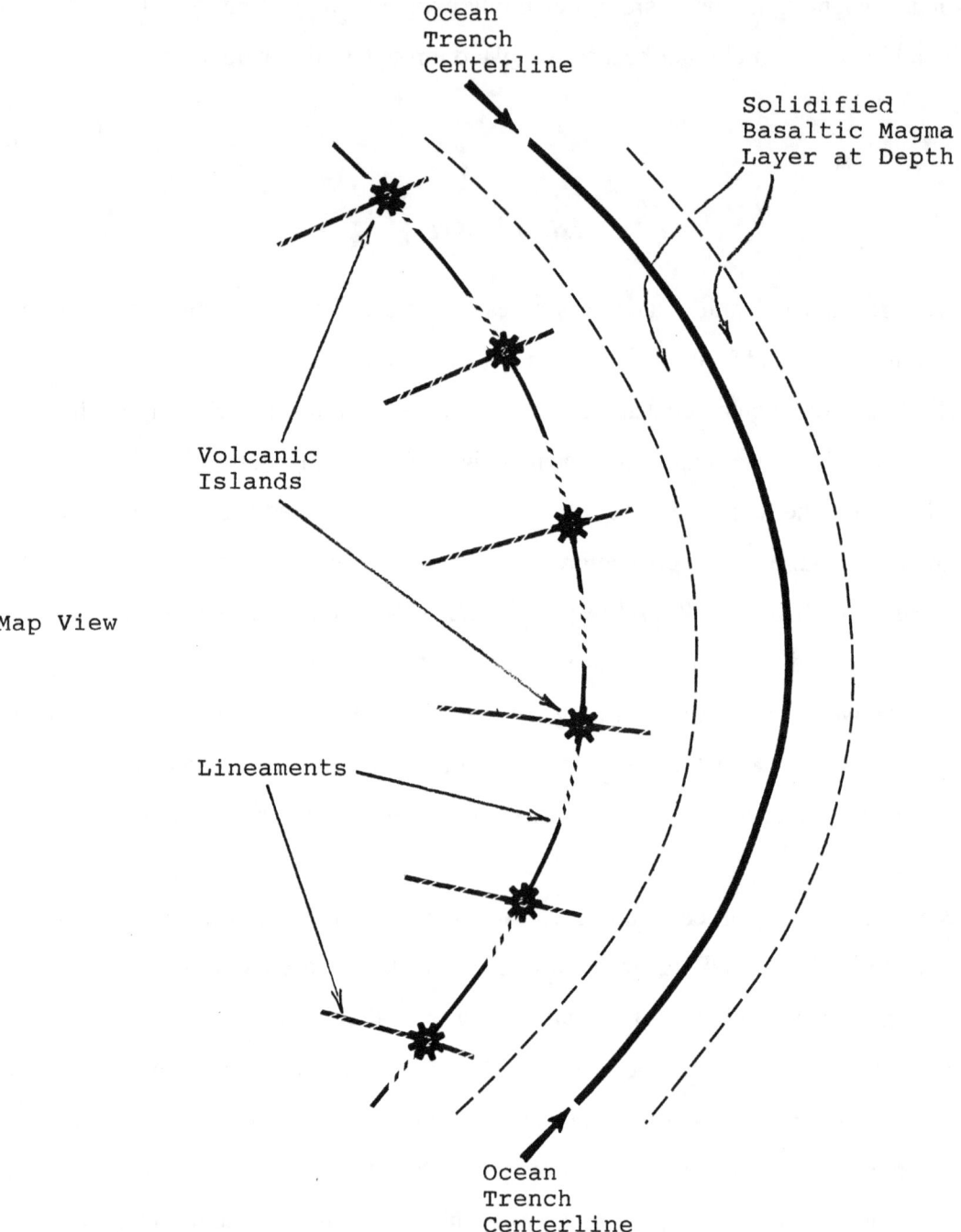

VOLCANIC SEAMOUNTS

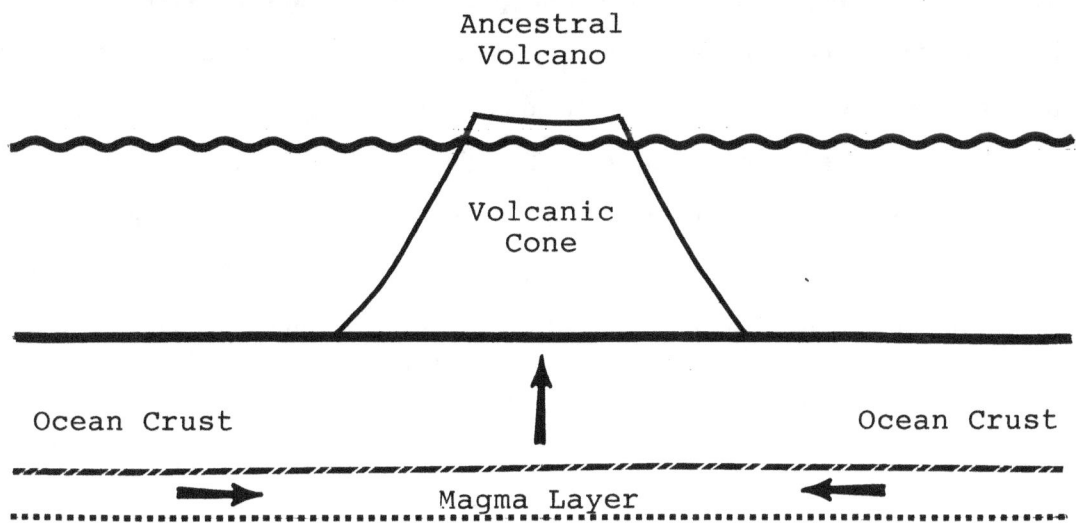

Vertically Exaggerated Cross Sections

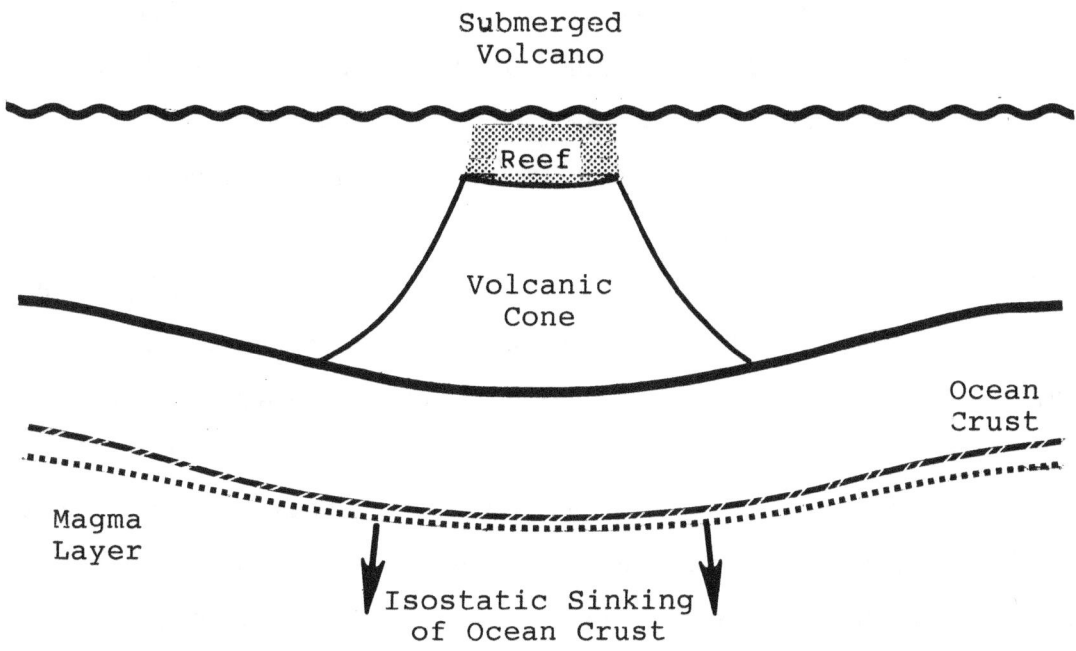

Volcanic eruptions occurred intermittently when magma pressures exceeded the sealing resistance of the widening fracture set above. **Lineament fracture intersections along the island-forming fracture set provided preferred locations for vertical conduits, explaining the somewhat uniform spacing of volcanic islands.** The transfer of material from the magma reservoir to the volcanic island above eventually caused the seafloor crust below to subside. Overburden subsidence tilted the surrounding ocean floor crust and redirected the basaltic magma flow away from the active volcanic conduit and toward an adjacent lineament intersection conduit. This process sequentially raised one volcanic island after another until the island chain was completed and basaltic magma depleted.

Pockmarks

One possible remnant of oceanic column tilting is the unusual V-shaped pits on the ocean floor, called "pockmarks". Pockmarks are described by several authors (King and MacLean 1970, pp. 3141-3148; Hovland and Judd, 1988, pp. 58-118; Kelly, et al., 1994, pp. 59-62; Paull, et al., 1995, pp. 89-92).

Pockmarks appear to be associated with gas and fluid escape from rocks below. Visualize collapse progression direction at a 45-degree angle to intersecting lineaments. The lower portion of each collapse block is pushed further in the direction of collapse advance, than the overlying crust adjacent to the sea floor. Lineament-bounded blocks are therefore tilted back, orienting the low corner of each block in the direction from which collapse had advanced.

Periodic lineament displacement opens fracture conduits for escaping fluids and gases. The intersection of lineaments provides the most favorable volatile escape conduits to the surface. Escaping fluids and gases rise toward the ocean floor through tilted lineament plane fissures that form the sides of the overhanging block. Their escape into the ocean waters above "flushes out" sediments along those lineament boundaries. That forms the characteristic "V-shaped" pockmark pits in the ocean floor, at the down-thrown surface corner of each tilted block. **Those "V-shaped" pockmarks point in the direction opposite to collapse advance and horizontal displacement of Earth's crust.**

POCKMARKS
MAP VIEW

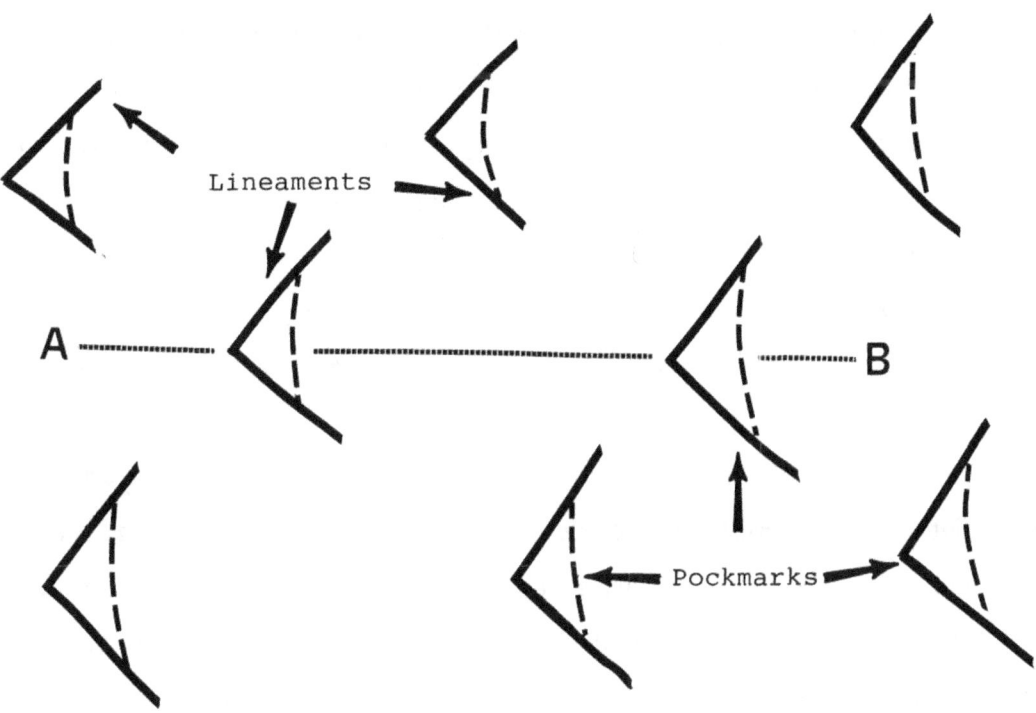

CROSS SECTION A - B
(Not to Scale)

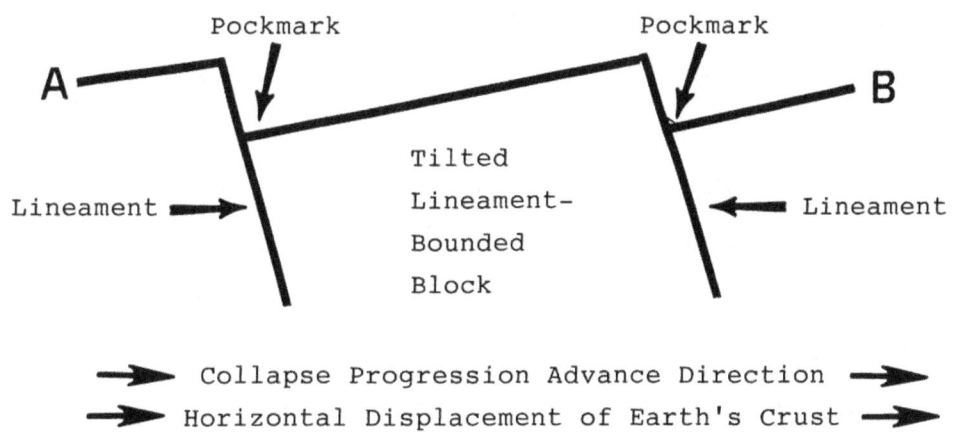

Seafloor Basalt Intrusion

Magma intrusion into the collapsing crust above was facilitated by the following geochemical process. *"The presence of water considerably lowers the melting temperatures of rocks. It also decreases magma viscosity and density, enhancing magma migration"* (Thompson, 1992, p. 295, also see Burnham, 1979, pp. 71-136). According to Mark Richards, University of California at Berkeley, *"hot spot volcanism could be triggered not by blobs of material that are anomalously hot rising through the mantle, but blobs of material that are anomalously wet."* Richards points out that the presence of volatiles, such as water in a mass of rock, would make it buoyant, causing it to rise. *"The presence of volatiles also lowers their melting point, so that when they rise to the surface, you get massive melting,"* says Richards (Bergeron, 1997, pp. 25-26).

During the Regressive Stage of the Worldwide Flood, ocean basin collapse crushed the porous rock matrix. Porous crust collapse released *pyromagma* volatiles including water and carbon dioxide from vugs in the freshly-shattered crust. Collapse-induced compression injected *basaltic magma* upward through the pulverized crust above. *Basaltic magma* viscosity was reduced upon contact with *pyromagma* volatiles. Overpressured *basaltic magma* freely flowed upward, melting and assimilating pyroclastic matrix minerals from the shattered crust until cooler temperatures adjacent to the ocean water above induced solidification.

*"The oceanic crust is commonly divided into three main layers: **layer 1** consists of ocean floor sediments and averages 0.5 km in thickness; **layer 2** consists largely of basalt and is 1.0 to 2.5 km thick; and **layer 3** is assumed to consist of gabbro and is about 5 km thick"* (Pratt, 2001, p. 17). Basalt and gabbro are both predominantly composed of plagioclase, pyroxene, and olivine, implying the same magma source. The uniformity of mineral components implies magma injection was too rapid to allow minerals to segregate by density or temperature of solidification. Basalt exhibits an aphanitic texture in which the crystalline components are too finely dispersed to be distinguished by the human eye. In contrast, gabbro slower cooling at greater depth allowed individual minerals to segregate separately at each mineral's solidification temperature, producing the granular gabbro texture. Therefore, cooling and solidification progressed very rapidly at shallow subsurface depths and progressed more slowly as depth increased. **Layers 2** and **3** are attributed to

Regression Stage collapse, intrusion, mineral comingling, and solidification. **Layer 1** is attributed to Flood and post-Flood sediment deposition.

From 1968 to 1977, Deep Sea Drilling Project boreholes encountered 19 sites, where the upper contact of the basalt with the sediments was baked (Meyerhoff, et al., 1992); implying those basalts are intrusive sills. *"In 33 cases, depositional contacts were observed, but the basalt sometimes contained sedimentary clasts, suggesting the potential for older sediments below"* (Pratt, 2001, p. 17; Hall and Robinson, 1979). **The *basalt* layer in the ocean crust suggests that magma flooding was once ocean-wide, and was accompanied by progressive crustal subsidence in large sectors of modern oceans** (see Keith, 1993; Beloussov, 1980). A sudden basaltic magma intrusion of ocean basins in the not-to-distant past is supported by the observation, that oceanic hot springs are usually much hotter than continental hot springs (see Kerr, Richard, 1987, p. 435).

Submarine Canyons

The Hudson River Canyon at the entrance to New York Harbor extends for one hundred miles into the modern day Atlantic Ocean (Velikovsky, 1955, p. 103). Submarine canyons, rivaling the length and depth of the Grand Canyon, extend far out onto the adjacent continental shelves, from the Amazon, Ganges, Congo, and Indus River systems (Brown, 1989, p. 63). Ice Age water captured by continental glaciers caused sea level to drop and expose some of the shallowest portions of submarine canyons. However, most submarine canyons extend outward along continental slopes to depths far below continental shelf exposure during the Ice Age. The size of these magnificent submarine canyons is far greater than channels cut by modern day river systems on coastal plains, implying their erosional origin was caused by a greater magnitude event.

Slump-induced turbidity currents are slow-flowing, seafloor-hugging, sediment-laden, density-flow currents. Turbidity currents slow where slope grade flattens, depositing sediments in submarine canyons on the continental rise, or as depositional fans on the abyssal plain below. Turbidity currents flow downslope toward and concentrate in submarine canyons. However, turbidity flows are incapable of eroding the deep submarine canyon channels described above. A better erosion mechanism is required.

SUBMARINE CANYON EROSION

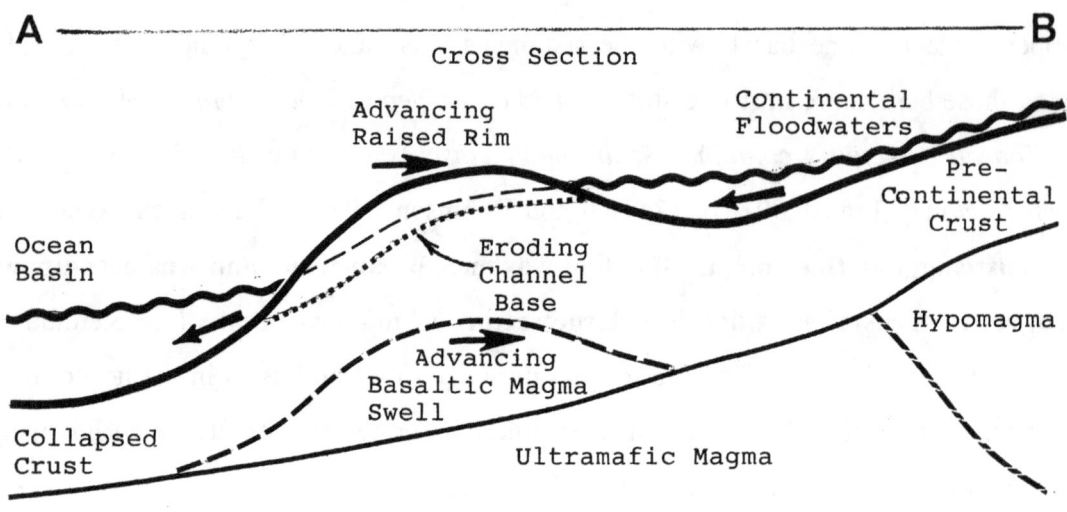

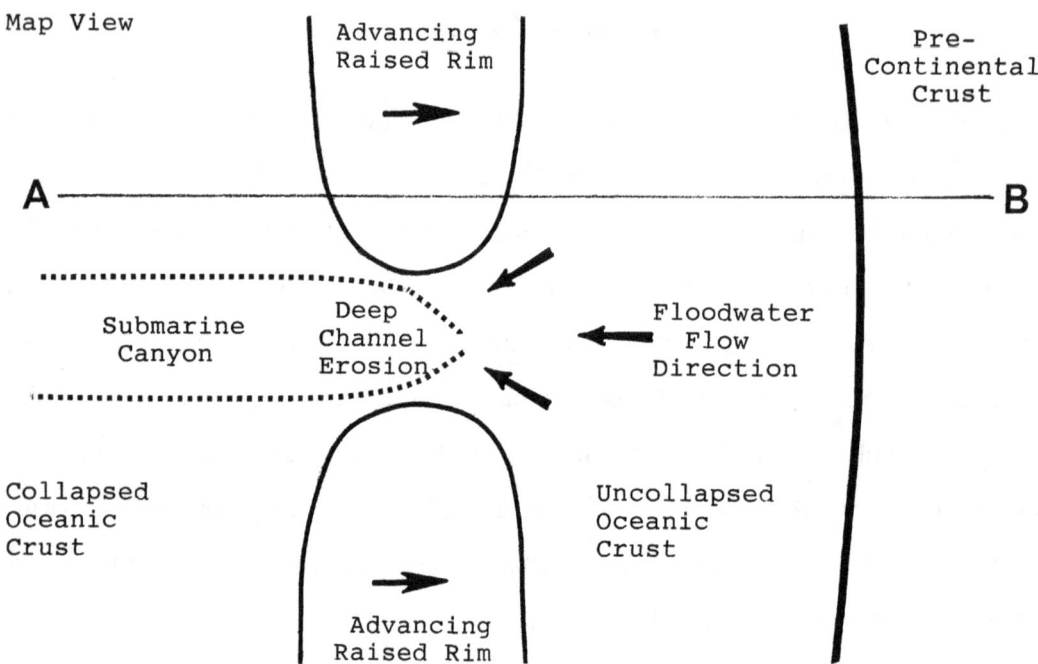

As the Worldwide Flood collapse of ocean basin blocks advanced, vertical compression generated a "rolling pin" effect. *Basaltic magma* was "squeezed" horizontally in front of the advancing seafloor collapse. That leading subsurface *basaltic magma* swell lifted the next contiguous uncollapsed block, causing a temporary upward flexure in Earth's surface immediately preceeding the ocean basin collapse that followed. However, the advancing rim slowed as it rose to higher elevation approaching the continental boundary (where Antediluvian crust thickens), thereby slowing the pace of collapse advance. Therefore, the raised *basaltic magma* swell of oceanic crust turned and aligned more parallel to the continental boundary as it approached the continent.

Floodwaters were rushing off continents toward collapsing ocean basins along future major river drainages. Meanwhile, the leading rim of elevated crust followed by collapsing ocean basin crust was advancing in the opposite direction toward the emerging continent. Floodwaters flowing off continents overflowed the lowest topographic elevation along the crest of the advancing raised rim of soon-to-be-collapsed pre-oceanic crust. A deep channel was carved into the ocean side of the advancing rim as continental floodwaters plunged down into the enlarging ocean basin. **Therefore, submarine canyons were progressively cut from an offshore position toward the continents on the backside of the advancing elevated rim.** Ocean basin collapse behind the advancing rim immediately dropped these submarine canyons far below modern sea level.

Raising Continental Rims

During the Regression Stage of the Worldwide Flood, ocean basin collapse advanced around the world, butting into the edge of emerging continents. Vertical porous crust collapse compression could not be directed downward into higher-density materials below and was diverted laterally. Lateral compression flattened and displaced collapsing oceanic crust in rolling pin fashion, expanding seafloor crust laterally. Laterally-expanding oceanic crust underthrust beneath adjacent lower-density, topographically-higher continental crust.

The underthrusting of oceanic crust produced an anomously thick vertical section of low-density crust, isostatically uplifting continental rim monoclines. Rising monocline edges were unsupported on the oceanic side. Continental boundaries commonly exhibit both compressional thrust faulting and tensional normal faulting. Thrust faulting was generateded by compressional ocean crust underthrusting. Tensional faulting was induced by sea floor collapse tilting of continental slope sediments. Tensional gravity slumping occurred as hanging wall blocks rotated in cross section, sliding down and out to sea. Those slide surfaces formed normal, seaward-dipping, listric faults along arcing planes that flatten with increasing depth.

CONTINENTAL RIM FORMATION

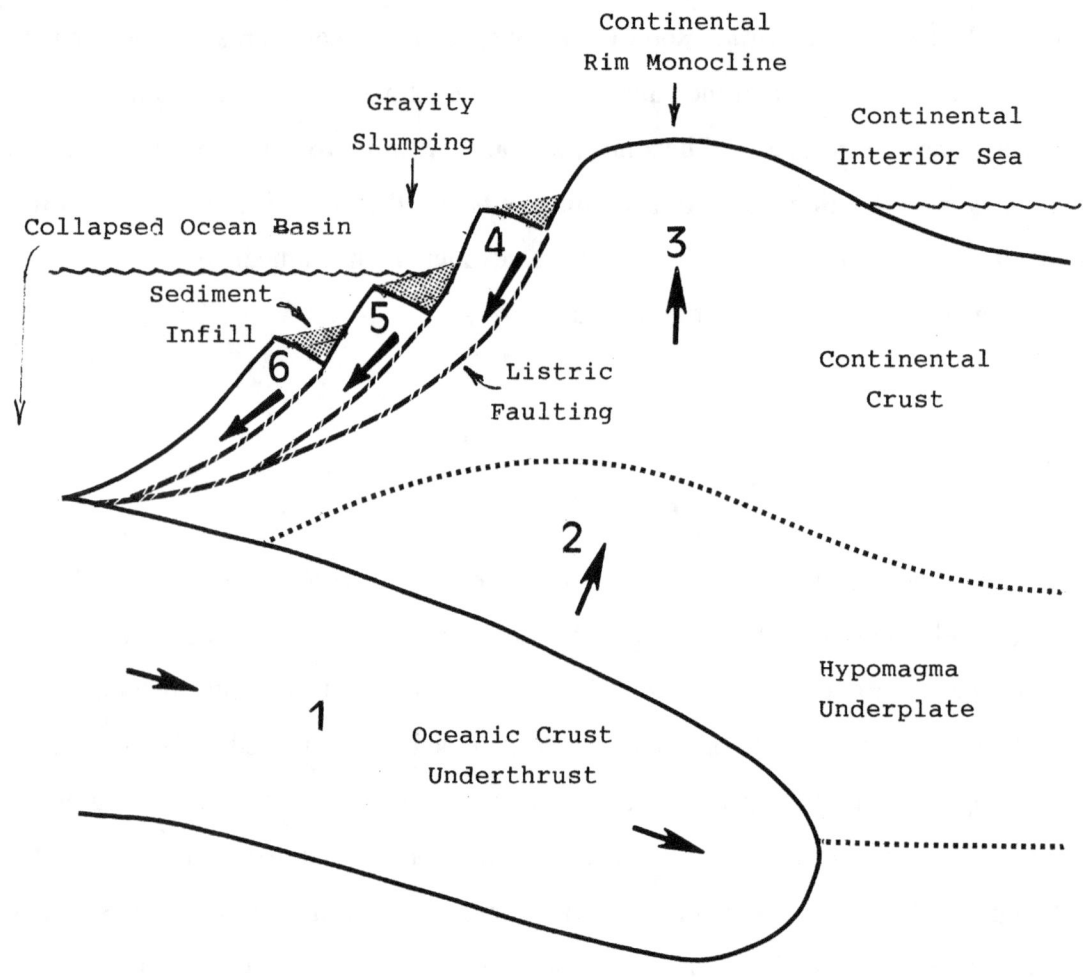

Vertically Exaggerated Cross Section

Coastal Monocline Uplift

A gently-dipping monoclinal rise or swell commonly exists inland and parallel to many continental coastlines. *"Antarctica and Greenland both have marginal swells. It is interesting that both have huge ice caps in the center and mountains around the rim. It is generally thought that the weight of the ice depresses the crust, and this is no doubt correct, but since many other continents have central depressions too, this might not be the whole answer"* (Ollier and Pain, 2000, p. 217).

Lester King is credited with first recognizing that the intersection of monoclines produces extra topographic uplift (Ollier and Pain, 2000, p. 193). Greater topographic uplift at monocline intersections indicates the compressional tectonic energy responsible for generating monoclines is converging from two ocean sides.

The following three examples portray continental rim monoclines that intersect at roughly 120 degrees. *"For example, in Australia the northerly monocline of New South Wales meets the east-west monocline of the Victorian Highlands, with Mt. Kossciusko in the corner – the highest mountain in Australia. In South Africa the Natal monocline meets the Cape Province monocline at the Compassberg, the highest point in Cape Province. Likewise, the East Brazil monocline intersects the Paraiba monocline at Pico da Bandiera, the highest point in Brazil"* (Ollier and Pain, 2000, pp. 193-194). **The convergence of lateral displacement generated by oceanic crust underthrusting accentuated isostatic uplift where continental rim monoclines converged.**

Earth's interior is slowly degassing and shrinking. The overlying crustal covering is similar to a man who looses weight and discovers his clothes are too big. So crustal overlaps are continually reactivated. Other weakened areas of crust slowly buckle up from horizontal compression like wrinkles on an oversized garmet. Slow degassing of Earth's interior and consequent lateral displacement of Earth's crust has been wrongly attributed to Plate Tectonics movement. However, the subsidence of ocean basins and the emergence of continents occurred rapidly during the Regression Stage of the Worldwide Flood year.

Mid-Atlantic Ridge

Coarse-grained sand from "beach-like" terraces on the Mid-Atlantic Ridge has been brought up from depths of two miles and three and one-half miles (Ewing, 1949, p. 613). *"According to Ewing, long flat stretches were detected 2 to 20 miles wide and hundreds of miles long. These beach-like areas were always covered with thick sediments, indicating a long period of deposition, although occasionally separated by mountainous "higher ground" exhibiting no such sediments ... Ewing observed that deep ocean basins never have thick sediments – which are the result of surf action and river deposition – it is actually shorelines that display thick sediments ... the deposits were found to be well-sorted by surf action into the usual pattern of shoreline beaches familiar to geologists (Miller & Scholten, 1966)"* (www.atlantisquest.com/Geology.html; also see Ewing, 1948).

The Swedish Museum of Natural History sponsored a Deep Sea Expedition that retrieved deep sea cores along the Mid-Atlantic Ridge containing fossils of terrestrial plants and a layer of fresh-water lake diatoms (Kolbe, R.W., 1957: Kolbe, R.W., 1958). These terrestrial evidences imply that large areas along the submerged Mid-Atlantic Ridge were subaerially exposed for several centuries following the Worldwide Flood, and subsided thereafter several kilometers to their present depth. Plato's writing about demise of Atlantis, which was located beyond the "Pillars of Hercules" (Strait of Gibraltar), probably refers to a cataclysmic sinking of the Mid-Atlantic Ridge.

Some of the most ancient maps show Atlantic islands that are submerged today. *"One is that big island on the Piri Re'is Map is located right over the Mid-Atlantic Ridge ... at the spot where two tiny islands, the Rocks of St. Peter and Paul, jut up above the sea, just north of the equator and about 700 miles east of the coast of Brazil"* (Hapgood, 1996, p. 65). The Piri Re'is Map of 1513 is probably the last of a succession of copies perhaps dating back to an original map made centuries or millennia earlier.

Horizontal compression of Earth's crust progressively increased during the Worldwide Flood Regression. Compression-driven collapse generated orthogonal seafloor topography, characteristic of the Atlantic region. At the end of the Worldwide Flood, Earth's interior was rapidly degassing through the oceanic crust. Earth's crustal shell became too large to fit uniformly around its shrinking interior.

MID-ATLANTIC RIDGE FORMATION

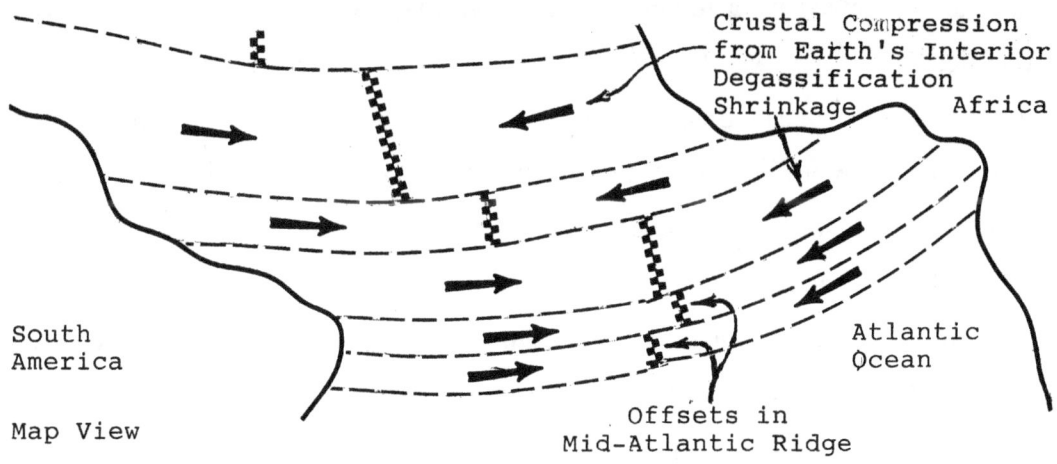

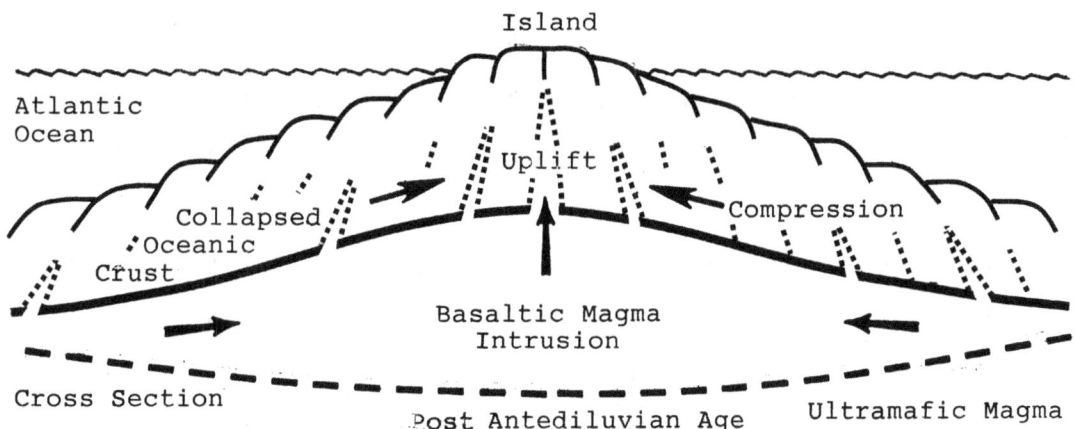

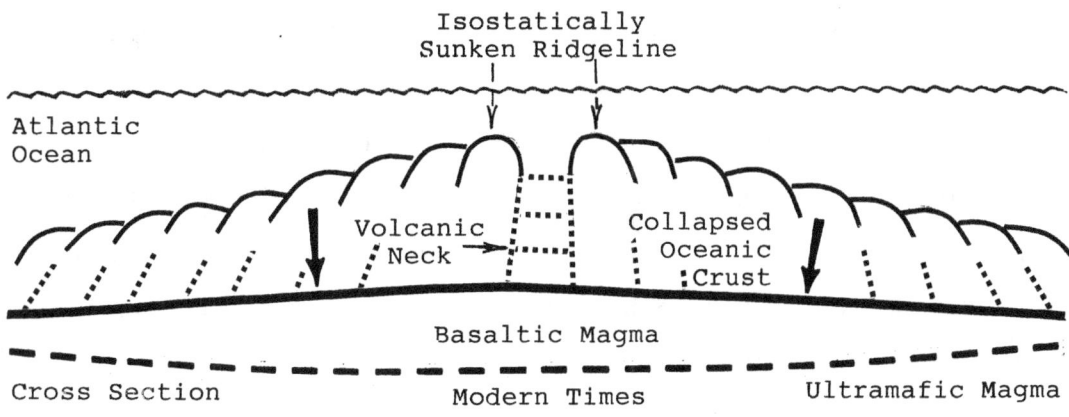

Ocean basin collapse generated vertical compression at depth and corresponding lateral seafloor extension. Adjacent continents were relatively stable platforms compared to collapsing Atlantic seafloor. Therefore, continental underthrusting by the Atlantic seafloor was immediately followed by a rebound reversal, shifting seafloor crust back toward the central Atlantic.

Because Atlantic Ocean Basin collapse progressed from south to north during the closing months of the Worldwide Flood, segments of the Mid-Atlantic Ridge were raised in sequence from south to north. Each sequentially-colliding segment of Atlantic seafloor crust is bordered by fracture zones, which are aligned perpendicular to the Mid-Atlantic Ridge. **The Mid-Atlantic Ridge buckled up where horizontal displacement of oceanic crust from the east collided midway with horizontal displacement of oceanic crust from the west, uplifting islands above the waves.**

The American and African/European continental rims are irregular (not parallel) in shape. That explains why the Mid-Atlantic Ridge is not continuous, but instead is characterized by many segmented offsets. Remnant *basaltic magma* flowed up dip beneath the tilting Atlantic crust and enlarged a magma chamber beneath the rising Mid-Atlantic Ridge.

Horizontal compression in Earth's crust slowly dissipates unless continuing stress is applied. Degasification of Earth's interior slowed, weakening horizontal compression. Isostatic forces prevailed as horizontal compression forces relaxed through time, causing the Mid-Atlantic Ridge to sink beneath the waves. The post-Flood subaerial exposure of much of the Mid-Atlantic Ridge was temporary. Brief periods of cataclysmic subsidence accompanied by catastrophic earthquakes occurred every few centuries, dropping most of the Mid-Atlantic Ridge beneath the waves during the Post-Flood Millennium.

A series of parallel faults were activated on both flanks as the Mid-Atlantic Ridge subsided. Periodic isostatic sinking of the Mid-Atlantic Ridge ruptured those faults and over-pressured the *basaltic magma* chamber below. Intermittent volcanic activity ensued, but tectonic activity decreased as the balance between isostatic subsidence and horizontal compression approached equilibrium.

Archean Cratons

Physical evidences of granite emplacement at depths of a few kilometers are difficult to explain. The quandary begins with a modern geothermal gradient of 20 degrees centigrade per kilometer in Earth's crust. That temperature gradient would not attain the 800 degree C temperatures needed to melt granitic materials above a depth of approximately 35 km (see Thompson, 1999, pp. 7-25). Granitic magmas are 10 to 1,000 times more viscous than basaltic magmas (Baker, E.R., 1996, pp. 126-134; Scalliet et al., 1996, pp. 27691-27699; Clemens and Petford, 1999, pp. 1057-1060). This flow resistance would severely inhibit rapid transport of granitic magma from 35 km to the surface. Cooling and solidification of granitic magmas should have occurred before reaching their present shallow position. However, there is little geological, geophysical, or geochemical evidence to mark the passage of such large volumes of granite magmas up through the crust (Clemens and Mawer, 1992, pp. 339-360; Clements et al., 1997, pp. 145-172; Petford, et al., 1993, pp. 845-848). Therefore, reasonably explaining the abundance of near-surface granite-gneiss complexes is problematic.

"Because we don't observe granites forming today, debate has raged for centuries as to how granites form" (Snelling, 2009, Vol. 2, p. 984). The deductive alternative is that overpressured low-viscosity hydrothermal fluids were rapidly transported from depths greater than 20 km and intruded into unconsolidated sediments at much shallower depths. Worldwide Flood events provide a reasonable tectonic mechanism. Oceanic crust underthrusting of emerging continents generated compression-induced thrust faulting. Thrust faulting initiated continental porous crust collapse, overpressuring the underlying crust-forming magma zone. Over-pressured *hypomagma* was forced upward through fault conduits and intruded collapsed porous crust. High temperatures and chemical components respectively melted and altered unconsolidated sediments that subsequently cooled and solidified into "Archean" granite-gneiss complexes.

Late Regression Stage collapse of porous crust continued into early post-Flood times at a dissipating rate of subsidence. Continuing crustal subsidence formed a synclinal trough along major fault zones. Subaqueous lavas were intermittently extruded along subsiding synclinal lows, forming greenstone belts. Greenstone belts are typically dominated by volcanic rocks including pillow lavas in the lower and middle portions, grading upward into sedimentary rocks at the top. *"The volcanic rocks of greenstone belts are typically greenish due to the abundance of the mineral chlorite, which formed during low-grade metamorphism ... The most common sedimentary rocks are successions of greywacke and argillite. Graywacke is a variety of sandstone containing abundant clay, and those in the greenstone belts are rich in volcanic rock fragments. Argillites are simply slightly metamorphosed mudrocks such as shale"* (Monroe and Wicander, 1994, pp. 524-525).

Regression Stage and early post-Flood tectonics initiated porous crust collapse and overpressured underlying *hypomagma*, generating prolific intrusion. That cataclysmic process established continental cratons underlying large portions of Earth's crust that are collectively referred to as "Archean" continental shields and platforms. **"Archean" granite-gneiss-schist complexes formed the most tectonically stable provinces on post-Flood continents.**

Worldwide Flood Heat Dissipation

The source of the "Fountains of the Great Deep" was the volatile-rich, crust-forming magma zone at a depth of approximately 25 kilometers (15½ miles). These fountains erupted from trenches concentrated in the Pacific Ocean basin. The fountains spouted pyroclastic volatiles and ash high into the atmosphere causing torrential rains during the initial 40 days of the Worldwide Flood. However, rising waters soon submerged the Fountains, causing rainfall to subside. Sealife adjacent to erupting trenches was scalded to death. However as floodwater spread across the land and mixed with Antediluvian surface waters, temperatures cooled and most sealife was not threatened during the Flood Inundation Stage.

Later, ocean basin collapse spread around the world during the Regressive Stage of the Worldwide Flood. Hot gas-rich volatiles including water and carbon dioxide were released into the floodwater above from vugs in collapsing porous pre-Oceanic crust. Most basaltic magma cooled and solidified before reaching the surface. However, at some locations basaltic magma was extruded onto the seafloor and along contentinental flanks as subaerial "Precambrian" lava flows.

The intensity of surface heat release was limited by relatively shallow sources above 25 kilometers. Preeruption pyroclastic temperatures are estimated at approximately 1030 degrees C (see Melson and Hopson, 1981, p. 645). Heat release was concentrated above collapsing ocean crust, which progressively shifted around the world during the course of the Flood Regression Stage. There were always plenty of areas on Earth's surface where sealife was not threatened.

David Vonderheide (pers. comm., 2011) describes heat dissipation away from oceanic sources into the atmosphere and from there into space as follows:

1.) Turbulent mixing of water near extruding vents.
2.) Mixing within the Flood ocean by convective cells hundreds of miles in diameter.
3.) Atmospheric convection of steam rising to form giant thunderstorms.
4.) Adiabatic expansion in rising columns of air.
5.) Cloud-making via release of latent heat of condensation.
6.) Radiating heat into space from extremely high cloud tops.
7.) Returning cold rain and hail from high altitudes to the surface, cooling air and surface water.
8.) Cooler water sinking by reverse convection.
9.) Hypercanes turning heat into storm kinetic energy (rotation) that is in turn dissipated by turbulent mixing of spinning atmospheric eddies, and ocean wave formation.

Worldwide extinction of marine life was never threatened because areas distant from excessive heat release remained habitable throughout the Worldwide Flood. Future continental shield areas experienced life-threatening heat around localized volcanic vents during the Inundation Stage. However, most pre-continental areas provided sanctuary for marine life during the Worldwide Flood. "Perched" continental interior seas supported marine lifeforms for centuries thereafter.

During the Post-Flood Millennium, oceans were warmed by gradually dissipating heat conduction from the collapsed oceanic crust. Warm oceans evaporated prolifically into the atmosphere during the Worldwide Flood and at progressively slower rates during the Post-Flood Millennium. The atmosphere correspondingly released heat into space.

Surface temperatures during the Worldwide Flood never threatened to sterilize the planet because:

1.) **The crust-forming magma zone sourcing the Fountains of the Great Deep was located at less than 26 kilometers, limiting initial temperatures.**

2.) **Volatile temperatures decreased as liquid volatiles expanded into gases approaching the surface.**

3.) **Expulsion of crust-forming volatiles was dispersed throughout the initial 150 days of the Worldwide Flood, steadily dispersing heat flow.**

4.) **Boiling above vents and evaporation from adjacent floodwaters steadily transferred heat into the atmosphere.**

5.) **Basalt intruding into collapsing porous pre-oceanic crust was sourced from less than 30 kilometers, limitting initial temperatures.**

6.) **Intruding basalt cooled as it ascended through collapsing porous crust. Most intruding basalt solidified before reaching the surface.**

7.) **Porous crust collapse and associated basaltic intrusion progressively advanced around the globe from day 150 to day 370 of the Worldwide Flood, steadily dispersing heat flow.**

8.) **Heat transfer from collapsed oceanic crust to oceans to atmosphere to space continued at a diminishing rate for millennia after the Worldwide Flood.**

POST-FLOOD CONTINENTAL CRUST COLLAPSE

Continental root-sourced *hypomagma* did not spread uniformly beneath the precontinental crust. The upward tilt of the crust toward the adjacent continental root prevented low-density pyromagma from being displaced uniformly downdip. Instead, extending "legs" of *hypomagma* flowed outward and around "bowl-shaped" lenses of *pyromagma*. Before and after the Worldwide Flood, *pyromagma* bodies trapped between laterally-expanding continental root underplates, solidified downward to form regions of deep porous crust. Pre-Oceanic porous crust collapsed almost entirely during the Regressive Stage of the Worldwide Flood. In contrast, porous crust collapse on emerging continental areas was limited to isolated areas weakened by Regression Stage Tectonics.

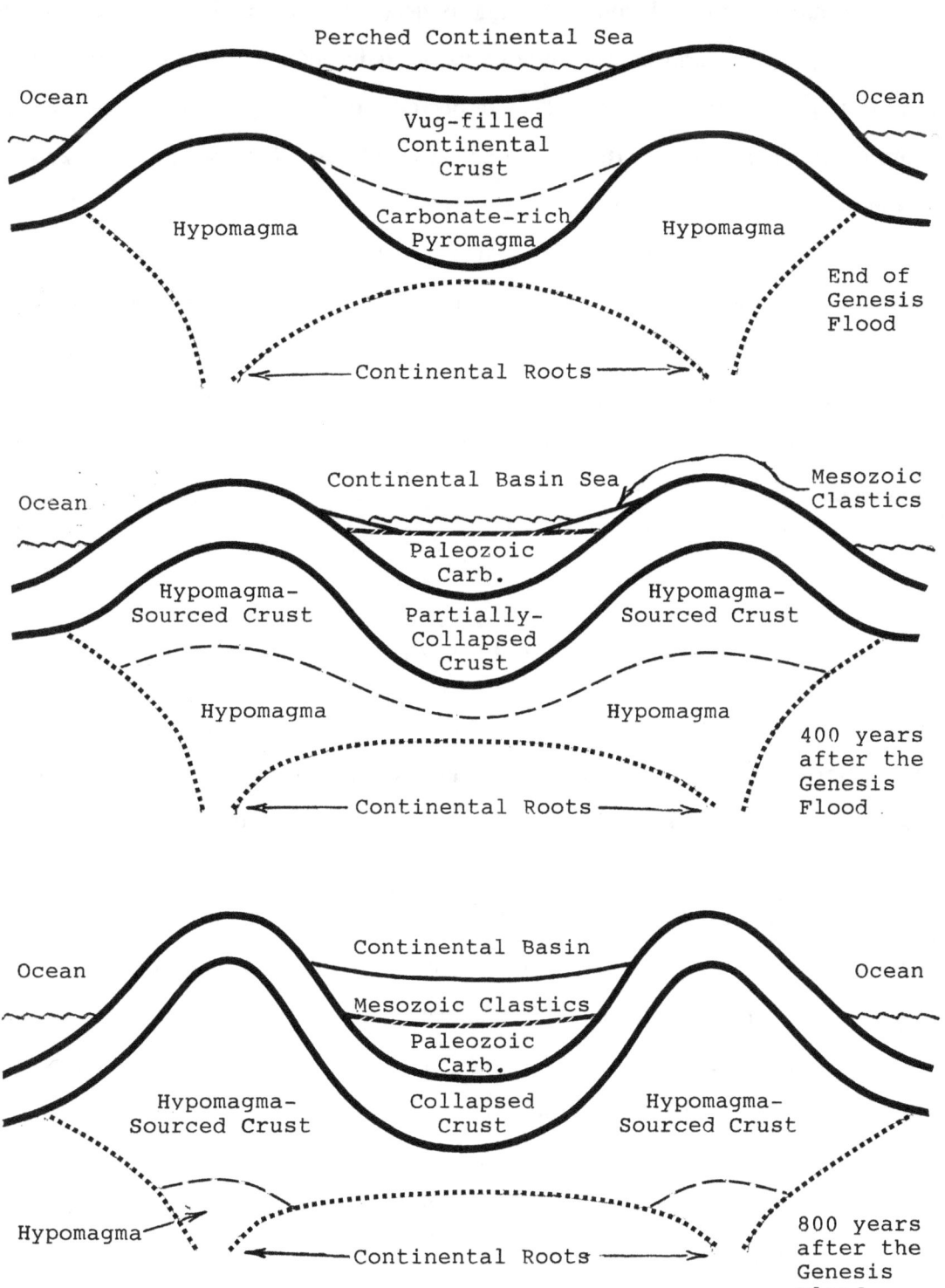

Worldwide Flood erosion removed large areas of impermeable primordial crust. Porous continental crust was now exposed to rock matrix decomposition by exposure to oxygen and groundwater circulation. Rock matrix decomposition allowed vug-encapsulated volatiles to escape to the surface. The loss of volatiles reduced overburden support, initiating collapse of the porous crust rock matrix. **The periodic deepening of exposure to oxygen and groundwater led to intermittenet porous crust collapse of deeper and deeper zones during the Post-Flood Millennium.**

Each crustal collapse event crushed volatile-filled vugs below, suddenly releasing a large volume of remnant *pyroclastic* liquids. These pyroclastic components contained an abundance of mineral salts, which were deadly to marine life. With each collapse eruption, mineral salts sank to the continental seafloor, depositing carbonate sedimentary formations. Paleozoic carbonate formations were predominantly sourced by this intermittent volatile-extrusion process.

Earthquakes

Modern earthquakes provide a subdued reminder of Earth's cataclysmic past. Collapse Tectonics implies earthquake movement should be predominantly vertical. What is observed? There is a predominance of vertical motion with earthquakes (Skobelin, et al., 1990, p. 26). Earthquakes occur when mass shifts closer to Earth's center and associated compression and sudden shear slippage emits shock waves.

Seismologists have difficulty accounting for deep earthquakes. At some undetermined depth (presumably about 30 kilometers), the pressure from the overburden of rock is so great that a crack cannot open up by itself. *"Instead of breaking suddenly by brittle fracture, rock at such pressures simply deforms when excessive shear is applied ... Yet earthquakes have been recorded from depths as much as 700 kilometers, and their seismographic "signatures" show that, like shallow earthquakes, they involve a sudden discontinuous fracture and slippage"* (Gold & Soter, 1980, pp. 157-158). Potential explanations for the cause of earthquakes below thirty kilometers include the following:

1.) Mineral lattice packing arrangement contraction or expansion in response to increasing or decreasing pressure.

2.) Liquid/gas conversion induced by pressure changes.

3.) Lower-density magma below exchanging positions with higher-density magma above.

The Rocky Mountain Arsenal northwest of Denver, Colorado, induced more than 700 shallow earthquakes from 1962 to 1965 by injecting contaminated waste water into fractured migmatitic gneiss at a depth of 11,975 to 12,045 feet. The maximum injection pressure was about 1,050 pounds per square inch, with an injection rate of 300 gallons per minute (Evans, 1966, p. 14). All earthquake epicenters (sources) were within seven miles of the injection well. As waste water was injected, fluid pressure increased, causing frictional resistance between blocks to decrease. Blocks are dually positioned by solid stress and the pressure of interstitial fluids. When fluid pressure approaches lithostatic pressure, shear stress approaches zero and the block will move freely. Movement of the block released elastic energy which generated earthquake waves (Evans, 1966, pp. 11-18). Therefore, porous crust collapse at depth may over-pressure connate fluids and trigger displacement, generating earthquake waves.

Let us now consider shallow (above 30 km) earthquakes generated by porous crust collapse. Liquefied gases in vugs are stable relative to the surrounding bedrock, so long as temperature and pressure remains constant. However, localized radioactive decay or heat generated by magnetic field decay may alter that temperature/pressure balance. Increasing temperature causes liquefied gases to expand and become slightly over-pressured relative to the surrounding bedrock. This overpressuring generates micro-fracturing (dilatancy) between vugs and establishes a permeable network. Hydrogen and other low-density molecules escape upward, increasing compression at the base of the micro-fracture network.

"At this stage the pores at the bottom of the domain will collapse and the pores at the top will be opened; the pore-space domain as a whole will migrate slowly upward. Because the average pressure of the gas decreases during the ascent, the volume of the pore-space domain must increase. This phenomenon may be the cause of the gradual rising of the surface (about 10 centimeters) that has been detected over a period of years preceding an earthquake" (Gold & Soter, 1980, p. 161).

If the porous bedrock matrix is extensive enough, the compression rate of the micro-fracture network suddenly causes the base to collapse. This produces dull "explosive" noises and generates seismic waves. Collapse injects liquefied gases upward into the overlying micro-fracture network. Increasing collapse-induced pressure accelerates the rate of upward expansion of the micro-fracture network above. Structural collapse at depth reduces support capacity of porous bedrock above. That explains why earthquake epicenters routinely rise from deeper to shallower zones as the overlying crust progressively collapses upward.

The micro-fracture network eventually breaches fracture conduits open to the surface usually at depths less than five kilometers. Liquefied gases race toward the surface, expanding into gas as pressure rapidly decreases. Rapid depressuring of the micro-fracture network below generates collapse and emits shallow seismic waves. Volatiles entering magma bodies reduce viscosity. The collapse of overlying bedrock generates compression, sometimes injecting magma into the strata above. Magma sometimes erupts at the surface, expelling volcanic ash or lava.

Earth's modern crust is riddled with fractures that allow *pyromagma* volatiles to excape to the surface. Consequently, *pyromagma*-rich volatiles have become depleted from subsurface magmas through time. The formation of new porous crust has been negligible since the Post-Flood Millennium. Most of the porous crust that formed so prolifically during the Antediluvian Age collapsed to form ocean basins during the Regression Stage of the Worldwide Flood. Continental porous crust collapse and resulting shallow earthquake activity was concentrated during the Post-Flood Millennium. Continental earthquake activity accelerated during the Post-Flood Millennium as porous crust intermittently collapsed downward to its base in most continental basins. Thereafter, shallow earthquake activity associated with collapse of remnant lenses of porous crust steadily decreased into modern times.

Earthquake data for comparison purposes only dates back to the 1930's. Austin and Strauss (1999, pp. 30-39) concluded *"1. a comparative excess of big earthquakes occurred in the first half of the century, and 2. An obvious decrease in the frequency of big earthquakes occurred since 1950."* Also, the data indicated no significant increase in the frequency or intensity of shallow earthquakes since 1977. **Earthquake data loosely supports the concept that Earth's crust is continuing to stabilize.**

Sumatra-Andaman Earthquake

The Sumatra-Andaman Earthquake (SAE) of December 26, 2004 launched a tsunami wave that roared onto the coastlines of eastern India and southern Asia, killing over 200,000 people. SAE ruptured along the curved Sunda Trench fault zone where ocean crust underthrusts adjacent continental crust to the northeast.

During the Regressive Stage of the Worldwide Flood, earthquake-prone Wadati-Benioff Zones formed along some contacts between pre-continental crust and pre-oceanic crust. Collapsing ocean basins generated outward horizontal displacement of Earth's ocean crust. Wherever ocean basin collapse expansion butted into continental crust, the oceanic crust was underthrust beneath adjacent lower-density continental crust.

Underthrust angles along slippage planes are widely varied according to the lateral approach angle of overlapping crust. "The (SAE) fault segment dip angles were approximated with the use of seismicity-based slab contours, which imply that the dip of the fault segments increases from south to north at" 12, 15, and 17.5 degrees (Ammon, et al., 2005, p. 1136-1137). Therefore, seafloor collapse advance approached the continental crust more perpendicular to the south end of the fault. However, the curvature of the boundary of continental crust was approached by collapsing crust at a more acute angle to the north, producing a more steeply-dipping fault plane.

Modern fault slippage along dipping Wadti-Benioff Earthquake Zones is attributed to a gradual contraction of Earth's interior. Low-density materials rise as high-density materials sink in a flowing medium. This relentless density segregation allows Earth's lower-density components to routinely escape into the atmosphere by volcanic activity and upward migration through the crust. This one-way gas and liquid expulsion process continually shrinks Earth's interior. Wadati-Benioff Earthquakes periodically occur when Earth's crust shifts along "overlapping" crustal boundaries.

Earth's shrinking interior gradually produces a smaller surface diameter. As Earth's diameter gradually decreases, less surface area is available for Earth's oceans. For example, SAE raised global sea level by about 0.1 mm. This sea level rise is popularly attributed to a net uplift in the Bay of Bengal and the Andaman Sea (Bilham, 2005, p. 1126). Although crustal overlap may cause isostatic uplift, this 0.1 mm sea level rise may be in part due to Wadati-Benioff-induced reduction in oceanic surface area.

Earth's rotation speed increases as higher-density materials concentrate closer to Earth's center. Consider an ice skater who starts a spin with arms and leg extended, then tucks her arms and leg close to her body to spin faster. The 9.1 SAE earthquake shortened the length of a day by 6.8 microseconds. The 8.8 Chile earthquake of February 27, 2010 shortened Earth days by 1.26 microseconds (Gross, 2010).

As ocean basins collapsed during the Worldwide Flood, Earth's surface area contracted. **Therefore, Earth's rotation speed was slightly slower during the Antediluvian Age.** A slower rotation speed produced fewer calendar days during each orbital revolution of Earth around the Sun. Early post-Flood cultures may have based their 360 day calendar year on Earth's slightly slower pre-Flood rotation rate.

New Madrid Earthquakes

Earthquakes are frequently associated with volcanic activity and crustal displacement along "continental plate boundaries". Those earthquakes are commonly attributed either to ascending magma or to displacement generated by adjacent continental plates shifting in different directions. However, the most severe earthquake sequence in United States history occurred far away from any volcanically active area and even further away from any continental plate boundary. The New Madrid (Missouri) Earthquakes of 1811 and 1812 provide compelling evidence of shallow earthquakes generated by porous crust collapse.

Garland Broadhead (1902, pp. 76-87) compiled the following eyewitness accounts at various locations and dates following the New Madrid Earthquake.

"The original focus of these concussions was the valley of the Mississippi, between New Madrid and Little Prairie ... But after the second year of their duration they seem to have ascended the Mississippi to the Ohio, then up that river 100 miles to the U. S. Saline ... The stronger shocks of this great series were felt in every part of the (continental) United States" (Dr. Drake).

"The earth began to totter and shake so that persons could neither stand nor walk" (Godfrey LeSieur). *"There were two classes of shocks: those in which the motion was horizontal, and those in which it was perpendicular ... the western sky was a continuous*

glare of vivid flashes of lightning and of repeated peals of subterranean thunder" (Timothy Flint).

"The ground rose and fell in successive furrows, like the waters of a lake ... The earth waved like a field of corn before the breeze" (J. J. Audubon). *"Then the earth was observed to roll in waves a few feet high with visible depressions between. By and by these swells burst, throwing up large volumes of water, sand and coal. Some was partly coated with what seemed to be sulphur. When the swells burst, fissures were left running in a northern and southern direction and parallel for miles ... The rumbling appeared to come from the west and travel east"* (Godfrey LeSieur).

"During all the hard shocks the earth seemed horribly torn to pieces ... Some of these fissures closed up immediately after they had vomited forth sand and water" (Eliza Bryan). *"A few persons fell into the chasms but were extricated"* (Timothy Flint). *"Fissures would be formed 600 to 700 feet long and 20 to 30 feet wide through which water and sand spouted 40 feet high. There issued no burning flames but flashes such as would result from an explosion of gas, or from passing electricity from cloud to cloud"* (J. W. Foster).

"There seemed to be a blowing out of the earth, bringing up coal, wood, sand, etc., accompanied with a roaring and whistling produced by the impetuosity of the air escaping from its confinement seemed to increase the horrid disorder of trees being blown up, cracked and split and falling by thousands at a time. The surface settled and a black liquid rose to the belly of the horses who stood motionless, struck with panic" (L. Binegler). *"The water thrown up during the eruption was luke warm"* (Godfrey LeSieur). *"After the severest shocks a dense black cloud of vapor over-shadowed the land. The sulphur gases escaping through the cracks, the air became tainted and the water for 150 miles was so impregnated as to render it unfit to use"* (Henry Howe).

"Many persons perished with their boats upon the river" (Timothy Flint). *"After daylight, as they were preparing to depart, a loud roaring was heard sounding like steam escaping from a boiler. This was accompanied by a violent agitation of the shores and tremendous boiling up of the waters in huge swells and tossing the boats so violently that the men with difficulty could keep upon their feet. The sandbars and points of islands gave way, swallowed up in the tumultuous bosom of the river, carrying down with them the cottonwood trees cracking and crashing, tossing their arms to and fro, as if sensible of their danger while they disappeared beneath the flood. The river which the day before was comparatively clear,*

being low, now changed to a reddish hue and became thick with mud thrown from the bottom, while the surface, lashed violently by the agitation of the earth beneath, was covered with foam, which gathering in masses the size of a barrel, floated along the trembling surface" (Hildreth & Linn).

"The fowls and beasts cried; trees fell; the Mississippi roared, the current for a few minutes was retrograde ... At first the Mississippi seemed to recede from its banks, its waters gathered up like mountains, leaving boats high upon the sands. The waters then moved inward with a front wall 15 to 20 feet perpendicular and tore the boats from their moorings and carried them up a creek closely packed for a quarter of a mile. The river fell as rapidly as it had risen, and receded within its banks with such violence that it took with it the grove of cottonwood trees which hedged its borders. They were broken off with such regularity that in some instances persons who had not witnessed the fact could with difficulty be persuaded that it was not the work of art. A great many fish were left upon the banks. The river was literally covered with the wrecks of boats" (Eliza Bryan).

"The site of the town (New Madrid) settled down fifteen feet ... the numerous large ponds which covered a large part of the country were nearly dried up. The beds of some of them were elevated above their former banks several feet ... A lake on the opposite side of the Mississippi has been formed over 100 miles long, six miles wide and ten to fifty feet deep" (Eliza Bryan). *"Reelfoot Lake in Obion Co., Tennessee, nearly 20 miles long and seven broad owes its origin to the sinking of the ground during this period"* (J. W. Foster).

"Besides long and narrow fissures, there were others of an oval or circular form making long and deep basins some 100 yards wide and deep enough to retain water in dry seasons. The damaged and uptorn country embraced an area of 150 miles in circumference" (Godfrey LeSieur).

"Afterwards the whole surface remained covered with holes which resembled so many craters of volcanoes surrounded with a ring of carbonized wood, and sand which rose for about seven feet. A few months after, these were sounded and found to exceed 20 feet in depth ... There seemed to be a tendency to carbonization in all vegetation soaking in the ponds, produced by these eruptions" (L. Binegler).

The above accounts provide an abundance of clues concerning the geologic mechanism that produces earthquakes. The rise and collapse of the ground surface is closely associated with the eruption of vast quantities of subterranean gases and liquids. *1.) How are those gases and liquids contained before an earthquake?* Vugs encased in surrounding rock matrix provide the most reasonable means of liquefied gas containment for an extended period of time. Otherwise, those gases and liquids gradually leak and escape at the surface. *2.) What geologic mechanism abruptly releases those gases under extreme pressure to the surface?* Porosity collapse crushes the vugs, releasing over-pressured liquefied gases. *3.) What causes sulfur-rich brine to be expelled at the surface?* Collapse releases hydrogen sulfide previously encapsulated in vugs. Fractures are generated above the collapsing zone, opening new conduits to the surface. *4.) How is the west to east "wave-like" motion of Earth's surface explained?* Collapse of a porous zone in Earth's crust induces additional pressure on adjacent uncollapsed crust. Therefore, the collapse of a porous zone shifts horizontally, squeezing expelled liquids ahead of the collapse front in advancing subsurface waves that lift Earth's surface above via the characteristic "wave-like" motion.

The New Madrid Earthquake was not unique in being located far from plate boundaries. *"On March 25, 1998, the largest earthquake ever recorded on the ocean floor struck inside the Antarctic plate, 350 kilometers from the nearest plate boundary"* (Monastersky, 1994, p. 155). Other powerful intraplate earthquakes include Lisbon, Portugal (1755) and Charleston, South Carolina (1886) (Brown, 2001, p. 112). **Shallow earthquakes far away from theorized plate boundaries are attributed to porous crust collapse.**

Mud Volcanoes

Magmatic volcanoes expel molten rock possibly sourced from Earth's mantle. In contrast, much cooler temperatures exhibited by mud volcanoes are more reasonably sourced by porous crust collapse events at depths above 25 km. There are approximately 700 active mud volcanoes worldwide. Clare Doyle of the BBC provides the following eyewitness account of the 10 October, 2001 Lokbatan eruption. *"There was a huge explosion, and a huge flame started coming from the hillside"* said one witness. *"It looked as though an animal was trying to emerge from the ground. The flame was unbelievably large – about 300 meters high. It was surrounded by dense, black smoke, and lots of mud was being thrown into the air. The largest flames burned for about five minutes. Then there was another huge explosion, and then the flames settled down to about 10 or 20 meteres (32 or 65 feet) high"* (Gallagher, 2003, p. 47).

The characteristic pattern of mud volcano activity often begins with an increasing amount of gas release at the volcanic vent a few days preceeding the eruption. This initial gas ejection constitutes a low-density, low-viscosity fraction of the volatiles sourcing mud volcano eruptions. *"... some components of gas (CO2, N2, He) and water (B, SO4, Cl) ... namely their anomalous increase during period of mud volcanoes activation which preceded seismic events in region ... The main gas component is methane (CH4) the content of which reaches up to 99% ..."* (Alyiev, 2004, pp. 199, 201). Flames associated with mud volcanoes are caused by combustion of escaping methane gas. *"One such famous gas seep is Yanardagh (Fire Mountain) on the Absheron Peninsula where a continuous fire burns along a hillside"* (Gallagher, 2003, p. 45).

Gas leakage from encapsulated vugs weakens the overburden support capacity of porous crust at depth. When a large enough lens of porous crust becomes unstable, collapse ensues generating the earthquake waves commonly associated with mud volcano eruptions. The sudden collapse of porous crust expels remaining volatiles as the vugs are crushed. Overpressured volatiles flow along the path of least resistance (usually faults and large fractures) to the surface. Upward-flowing diapirs form where fault and fracture conduits are insufficient to transport temporarily overpressured flow. As volatiles approach the surface, decreasing pressure causes liquid volatiles expand into gas. Upward volatile flow dislodges and assimilates adjacent sediments into the flow before erupting explosively, spewing silt and clay muds and lanching breccia missiles.

Mud volcanoes form cone-shaped deposits and surface mudflows. However, because the extruded materials are unconsolidated or poorly-consolidated, they are subject to rapid post-deposition erosion by wind, rain, and surface flow. These eroded materials are commonly transported downslope across subaerial environments and redeposited across the flank of subsiding basins. Consequently, evidences of mud volcanoes deep in the Rock Record are rare due to rapid post-deposition erosion.

Submarine mud volcanoes disperse solids and precipitate components throughout the hosting inland sea. The thickest accumulations collect where the water is deepest as fine-grained materials settle out of suspension during the months following an eruption. Shallow water mud volcanoes form temporary islands that quickly erode by wave action. Rapidly-accumulating mud volcano components are derived from the mineral content of 1) escaping volatiles expelled by collapsing porous crust, 2) materials accumulated during subsurface transport, and 3) dislodged surface materials. Established ecosystems are routinely overflowed and buried at the base of each mud volcano event.

Mineral components vary according to initial volatile content. For example, Paleozoic carbonate formations were primarily sourced by carbonate volatiles. In contrast, Mesozoic and Cenozoic shale formations were primarily sourced by volcanic ash components and small grain-sized components eroded and transported by surface flow. Clastic contributions progressively increased with proximity to uplifting areas approaching the end of the Post-Flood Millennium. Rapid burial of ecosystem materials on Earth's surface prevents premature organic decomposition. Increasing heat and pressure associated with increasing depth of burial facilitates petroleum generation from rapidly-buried organic materials.

Mud volcanoes are linked directly to areas of tectonic subsidence and deep sediment accumulation (Aliyev, 2004, pp. 192-193). Ash and volatiles squeezed out of collapsing porous crust are expelled at the surface. Collapse at depth is directly responsible for post-eruption subsidence of the surface above. Crater migration is associated with large mud volcanoes at Lockbatan, Bozdag-Gobu, Cheildag, Nardaranakhtarma and Airanteken (Aliyev, 2004, p. 198). The surface expression of the crater shifts laterally when porous crust collapse below shifts laterally. However, the surface position of the vent may remain in place because the fault or fracture conduit transporting volatiles to the surface remains in place.

Two deep seismic reflection profiles in the South Caspian Basin reveal typical sedimentary formation cross sections featuring distinctive, closely-spaced, sub-parallel reflectors approaching a horizontal orientation from the surface down to a depth averaging 7.5 kilometers (see Knapp, et al., 2004, p. 1077). This near-surface "productive zone" has been penetrated by many petroleum wells that verify the presence of Cenozoic sedimentary formations. *"... early studies suggested that the South Caspian crust is composed of an upper sedimentary layer, with a mean compressional wave velocity (Vp) of 3.5-4.0 km/s, and a lower oceanic ('basaltic') layer with Vp=6.6 km/s"* (Knapp, et al., 2004, p. 1074).

Seismic reflections occur when seismic waves traveling through a slower velocity material bounce off a denser velocity material below. *"The bright reflection at 26-28 km depth is interpreted as the basement/cover contact, and the underlying ~10 km portion of the section is interpreted as crystalline basement"* (Knapp, et al., 2004, p. 1073).

However, the slow velocity zone between 7.5 km and 26-28 km does not conform to normal seismic signatures of sedimentary strata. *"Further investigation revealed a non-uniform pattern of seismic wave propagation across mud volcanoes. When traveling through such a zone, the energy of elastic waves is attenuated, reflected waves almost completely disappear, and low frequency components dominate the wave pattern. The quality of the related seismic records dramatically deteriorates. A buried mud volcano and its breccia neither reflect nor refract seismic waves at the stratigraphic depth of such mud structure roots"* (Kadirov, et al., 2004, pp. 226-227). *"... despite the relatively large amount of earlier geophysical investigations of the South Caspian Sea region, the thickness and structural style of deformation of the sedimentary section as well as the crustal thickness and affinity remain equivocal"* (Knapp, et al., 2004, p. 1074).

So what is this mysterious anomalous slow velocity material residing between 7.5 km and 26-28 km? Seismic waves travel faster through denser material and slower through less dense material. *"The density of sedimentary layers depends upon composition and varies as follows: 1) for clays, 2.3-2.6 g/cm3; 2) for sandstones; 2.5-2.6 g/cm3; 3) for limestone, 2.6-2.9 g/cm3. In the zones of regional consolidation the clayey series is least dense, up to 0.2 g/cm3 less differentially"* (Kadirov, et al., 2004, p. 238). Those authors attributed this anomalously low clayey series density to thermal expansion. However, an alternative explanation would be anomalously high porosity sustained by upward migrating volatile liquids expanding into gases as pressures decrease.

If the disruption of seismic information between 7.5 km and 26-28 km is due to widespread, rapidly upwelling diapirs, then; 1) Why have mud volcano eruptions failed to flush the petroleum fields above? 2) Why do mud volcano vents remain in place when earthquake epicenters and crater subsidence migrate laterally? and 3) Why are listric faults mapped at the 10-20 km depth region where distinctive sedimentary formation reflectors are not observed? The implication is that the slow velocity zone between 7.5 km and 26-28 km is not composed of typical sedimentary formations.

Porous (vug-filled) crust is less dense than most sedimentary formations. The density of collapsed porous crust is expected to mimic the denser portion of the range of sedimentary formation densities. Porous crust is less dense than collapsed crust, so seismic reflections fail to rebound from the collapsed crust/porous crust contact. Therefore, compressional wave velocity measurements combine travel time through both porous and collapsed crust. The anticipated combined wave velocity travel time through the 7.5 km to 26-28 km zone provides a more viable explanation for the observed lower density discrepancy of 0.2 g/cm3 as acknowledged by Kadirov et al., (p. 238) above. **In conclusion, the intermittent collapse of porous crust provides the most comprehensive mechanism explaning mud volcano behavior.**

COLLAPSED POROUS CRUST EVIDENCE

Pyromagma-sourced porous crust contains no legitimate fossils because it solidified at kilometers of depth. The stratigraphic position of porous crust collapse must be "sandwiched" between underlying igneous and metamorphic basement bedrock and overlying sedimentary formations. The vug-filled *pyroclastic* porous crust mineral suite was only stable while the crust above acted as a permeability barrier.

Many *pyroclastic* minerals decompose and erode rapidly on exposed surfaces. Alkaline and acidic components which are present during volcanic extrusion commonly experience rapid alteration and dissolution upon exposure to oxygen and groundwater. Regarding Mt. St. Helens eruption pyroclastics, *"the deposits emplaced in 1980 are undergoing alteration as they cool and react with surface waters"* (Dethier, D., et al., 1981, p. 663).

As porous crust intermittently collapsed, groundwater altered and dissolved *pyroclastic* minerals. Dissolved minerals were transported away by groundwater toward oceans, leaving insoluble residue minerals behind. Subsurface dissolution, dispersal, diagenesis, metamorphic alteration, and amalgamation with intruding silicate magmas produced mineral suites quite different from the original porous crust. Modern porous crust remnants are probably limited to depths below seven kilometers due to circulating groundwater alteration and dissolution.

The optimum area to explore for remnant porous crust is near earthquake sites far from continental boundaries. In the United States, remnant lenses of uncollapsed porous crust are probably present at depth in the New Madrid Earthquake area. In the New Madrid area, the deepest earthquake epicenters (presumably representing the top of remnant porous crust) are probably located at approximately 18 km of depth. Earth's deepest drill hole in the Russian Kola Penninsula reached a depth of over 12 km (Kerr, 1989, p. 468). Consequently, remnant uncollapsed porous crust is typically encountered deeper than modern drilling technology has penetrated. If a remnant collapsed or uncollapsed porous crust lens is penetrated by well drilling, those materials would probably be wrongly interpreted from drill samples and wireline logs either as a porous carbonate precipitate formation or as mineral residue, expelled from a solidifying batholith. The seismic signature of remnant porous crust would be a slow-velocity zone with a distinctive crystalline basement reflector at its base.

Porous crust components have been redistributed into Earth's sedimentary, metamorphic, and igneous formations extending from Precambrian to Quaternary. Other porous crust components now occupy Earth's soils, oceans, and atmosphere. **It is therefore unreasonable to expect that a large volume of remnant porous crust would have been preserved into modern times in easily recognizable unaltered structural and mineralogic form.** However, the mineral residue of porous crust should be readily identifiable.

Precambrian Phosphorites

Precambrian phosphorites are a *pyroclastic* mineral suite residue derived from collapsed porous crust. Precambrian phosphorites are composed of phosphorous-rich apatite family minerals, dolomite, calcite, quartz, and assorted minor minerals. Precambrian Phosphorites are sandwiched above Precambrian crystalline basement and below Phanerozoic sedimentary formations – the anticipated stratigraphic position of porous crust residue.

We must first distinguish between Precambrian and Phanerozoic phosphorites. Precambrian phosphorites typically contain much higher concentrations of phosphorous than Phanerozoic phosphorites. Phanerozoic phosphorites were partially sourced by Precambrian phosphorites after being eroded and mixed with other sedimentary materials containing fossils. Phanerozoic phosphorites are characterized by abundant biologic evidences. *"A long-standing problem in Precambrian paleobiology has been why calcifying cyanobacteria are so rare, compared with their massive occurance in the Cambrian"* (Bengtson, et al., 2009, p. 5; from Riding R. (1994) pp. 426-438). **In contrast, Precambrian phosphorite fossil evidences are doubtful at best, implying these materials are comprised of collapsed porous crust residue.**

Precambrian Pseudomicrofossils

Body fossils are easy to recognize as biogenic because of their complex structure. However, microfossils are so tiny that diagenetic artifacts and metamorphic textures commonly mimic legitimate microfossil appearances. **Precambrian phosphorites have usually been altered by metamorphic processes, making microfossil identification extremely speculative.**

"Microstructures in the ~3.5 Gyr Apex Chert Formation were initially described as the oldest bacterial fossils on Earth over 20 years ago ... New samples were collected from the original fossil locality during the Geological Survey of Western Australia's (GSWA) field trip to the Pilbara Craton. Within the thick sections ... are features similar to the previously described microfossils. Raman spectroscopy was used to identify the fracture fill material on the thin and thick sections. The representative spectra acquired from the dark material in the thin and thick sections show diagnostic vibrational modes ... indicative of haematite ... our in situ Raman spectroscopic measurements demonstrate that the carbonaceous material is found in association with the matrix, not the opaque microstructures. Although the Apex Chert microstructures have many of the widely accepted characteristics of bona fide fossil microbes, these structures do not seem to be biogenic, nor do they resemble iron-stained or iron-replaced microfossils" (Marshall, et al., 2011, pp. 1-3). (Haematite a.k.a. "hematite" is an oxidized iron mineral.)

"Bacteria are basically little bags of goo, and they're not easily fossilized, said Alison Olcott Marshall. The idea that you would have this tiny bacterium preserved for 3.5 billion years is not very likely to happen. The second problem is that they are morphologically so simple – theyt're just circles and rods. There are lots of things in nature that make circles and rods" (Lynch, Brendan M., 2011, p. 1). In conclusion, supposed microfossils identified in Precambrian phosphorites are probably abiogenic.

Precambrian Stromatolites

Some Precambrian phosphorite zones contain wavy lamination "textures" or "fabrics" commonly shaped as domes, columns, and cones, which are typically millimeters to decimeters in size. Debate has raged for decades as to whether these laminated features were deposited as biogenic or abiogenic sediments before being metamorphosed. Regarding Precambrian strata, Awramik and Grey (2005, p. 1) state, *"No single feature or line of evidence has yet been found that can unequivocally indicate a biogenic nature for a stromatolite."*

Microscopic structures sometimes advocated to be stromatolite microfossils have potential alternative abiogenic causes. For example, *"The filamentous structures closely resemble fossil structures attributed to calcified filamentous cyanobacteria ..."* and *"The tubes resemble the concatenated cell walls of modern filamentous algae such as Spirogyra."* On the other hand, *"Spherulitic growth of the phosphate commonly produces cell-like structures that interfere with the original morphologies"* and *"... the embryo-like globules are not interpreted as fossils at all but as former gas bubbles ..."* (Bengtson, et al., 2009, pp. 1,2,4). Porous crust collapse did not expel all of the volatiles encapsulated in vugs. The so-called "gas bubbles" referenced above are remnant volatile-filled voids that survived collapse as residual textural features in porous rock.

Precambrian phosphorites are sometimes characterized by concentrically laminated calcareous masses, with irregular to columnar and hemispheroidal shapes. These textural features may be mistakenly interpreted as the depositional product of calcareous algae deposition known as "stromatolites". However, in situ diagenic mineralization associated with porous crust shatter, collapse, and expulsion of encapsulated volatiles could also generate features resembling the morphologic density and textural fabric of stromatolites.

Concentric laminations and irregular shapes in Precambrian phosphorite closely resemble the underlying high-grade metamorphic crystalline basement bedrock. In contrast, sedimentary Phanerozoic stromatolites typically show poor or no lamination. Consequently, using poorly-laminated Phanerozoic stromatolite analogs to interpret distinctive Precambrian phosphorite laminations limits that interpretation to an extremely low confidence level.

Phosphatic stromatolites are rare in Phanerozoic rocks. *"Phosphatic stromatolites had previously only been found in Proterozoic rocks"* (Southgate, 1980, pp. 395-397). These Australian Cambrian stromatolites were apparently deposited in a restricted evaporative basin rich in soda brine waters. Phosphate-rich apatite minerals do not precipitate directly from modern ocean water because the phosphorous concentration of 0.6 ppm (from Kranskopf, 1967) is too low. The rarity of phosphorous in Phanerozoic stromatolites is problematic because an anomalous source of prolific Precambrian phosphotization must be identified.

Finally, stromatolite deposition by calcareous algae is a slow process. That process occurs in warm shallow waters, with minimal sediment influx from outside source areas. A continuing outside source of phosphorous must be sustained for the duration of Precambrian phosphorite deposition. It would be difficult to imagine that a specific depositional environment could be sustained by gradual subsidence for long enough to accumulate more than a few tens of meters of stromatolite thickness. However, Precambrian phosphorites are commonly much thicker. An example of Proterozoic rocks rich in stromatolite-like features is the 4,500 meter thick sequence in the Peking area, China (Einsele, 2000, p. 288.) However, that broad stromatolite-like thickness is more reasonably explained as the residue of collapsed porous crust.

The preponderance of evidence favors the geomorphic interpretation that Precambrian phosphorites are the residue of collapsed porous crust. However, if any definitive fossil evidence can be found, that stratigraphic position and the overlying materials would be discredited as residue of collapsed porous crust.

Precambrian Metamorphic Mineral Fabric

Wavy laminations in crystalline basement gneiss are traditionally assumed to represent high-grade metamorphism of thinly-layered sedimentary materials. Igneous origins were not considered because millimeter-thick wavy laminations were not observed in batholith intrusives. However, Precambrian metamorphic mineral fabrics are better explained by diagenesis associated with porous crust collapse.

High-grade metamorphic crystalline gneiss is commonly characterized by varicolored mineralogies, which in cross section exhibit long, narrow subparallel stripes. The characteristic broad sweeping folds and wavy laminations are distorted in comparison to typical planar sedimentary depositional features. Therefore, layering contortions occurred prior to final consolidation into rock. That distorting process generated bending and folding instead of fracturing and faulting. Horizontal compression causing that bending and folding is traditionally attributed to pre-consolidation compression flow.

Horizontal compression provides a reasonable explanation for broad sweeping folds, which may measure hundreds of meters from crest to trough. However, wavy laminations as measured in millimeters or centimeters from crest to trough and laid out in series is problematic. Horizontal compression would squash and thicken unconsolidated sedimentary layers instead of wrinkling layers to form a well-defined series of thinly-rippled laminations. That begs for a more comprehensive hypothesis explaining the range of characteristics exhibited by Precambrian metamorphic mineral fabrics.

As an alternative interpretation, consider the viability of a crystal melt mush that solidified into a porous crust. The porous crust rock matrix was composed of an abundance of pyroclastic minerals, with volatiles encapsulated in tiny vugs. Porous crust stability depended upon the absence of exposure to circulating groundwater convection weathering and rock matrix deterioration. Pyroclastic mineral decomposition released volatiles, reduced overburden support, and caused that porous crust to collapse.

Collapsing vugs were squashed by overburden pressure from above. Rock matrix between horizontally-positioned vugs was crushed and shifted laterally via dilatancy to help fill vug space from the sides. Collapse shock pressure and friction caused low solidification temperature minerals to segregate and flow plastically along laterally-oriented planes interconnecting collapsing vugs. Horizontally-oriented mineral laminations aligned perpendicular to vertical collapse compression.

Porous crust collapse extended outward from its point of origin. Large domain collapse events shifted laterally like an advancing wave, compressing thick zones across large areas. The resulting mineral fabric is characterized by broad sweeping folds. However, major collapse events also typically left remnant lenses of porous crust within those recently collapsed zones. Subsequent collapse of remnant porous lenses progressed intermittently centimeters at a time, producing the characteristic wavy laminated mineral fabric. Intermittent small domain collapse events of thin zones advanced much more slowly and laterally in "spider-leg" fashion. The extending "spider-legs" formed lamination wave troughs as they "bottomed-out" first. Subsequent collapse of adjacent materials between the "spider-legs" did not sink as far, forming lamination wave crests. Domes, columns, cones, hemispheres, and irregular shapes exhibiting wavy laminations are attributed to isolated porous crust that collapsed following larger scale collapse events. For example, a late collapsing lens would progressively deteriorate and collapse from the flanks toward the center to form a dome-shaped feature.

Because the exponential difference in scale between broad sweeping folds and wavy laminations is explained by collapsing domain thickness and variable rates, porous crust collapse provides a more reasonable interpretation than metamorphic alteration of finely-layered sediments. **Therefore, the characteristic wavy structure of Precambrian gneiss is attributed to localized metamorphic alteration processes induced by porous crust domain collapse.**

Metasomatic Alteration

Plutonic rocks that cool and solidify from magma at their melting point are termed "magmatic". In contrast, "metasomatic" refers to a metamorphic process in which one mineral or mineral assemblage is replaced by another of different composition without melting. This chemical alteration occurs when hydrothermal fluids migrate through intergranular and fracture porosity. Migrating hydrothermal fluids deposit some minerals and remove other minerals from the surrounding bedrock. Metasomatism may be facilitated by altering pressures caused by tectonic compression or overburden removal. *"In most places, metasomatism is so complete that it is a problem to find remnants of the original rock which are undeformed and unreplaced"* (Collins, 1997, p. 13). Alteration of collapsed porous crust was most extensive where metasomatic alteration occurred.

Metasomatic alteration occurs well below the 600 degree Centigrade melting point of most mineral assemblages. *"The extent depends upon the degree of deformational movements, crushing, and availability of K, not on some arbitrary number ... During the recrystallization to produce these new metamorphic minerals, generally any pore spaces are eliminate ..."* (Collins, 1997, pp. 1, 13). Therefore, the original porous texture and mineral composition of collapsed Antediluvian crust was intensely altered as it collapsed and remineralized to form crystalline gneiss.

Metasomatic deformation was tectonically activated by (1) collapsing porous crust compression, (2) compression sourced by the uplift of adjacent mountain ranges, (3) decreasing pressure due to overburden removal, and (4) isostatic uplift generated by deeper continental root contributions. *"Where continued stresses are applied while metasomatism is occurring, foliation is enhanced, but where the rocks are cataclasticly broken but then are no longer under stress, the recrystallization that occure during the replacement process will eliminate any foliation and produce a massive-appearing rock"* (Collins, 1997, p. 13). Therefore, granitic textures can also be produced by intense metasomatism of collapsed porous crust.

"But the process does not end here. Many granitic rocks formed by metasomatic processes below melting temperatures may later melt because (1) volumes of introduced hydrous fluids, enabling metasomatism, can increase to the point that the system is saturated with water, (2) the temperature rises above eutectic conditions, or (3) the pressure becomes less because the metasomatic granite body has been mobilized to rise to higher levels in the crust" (Collins, 1997, p. 12). Magmatic penetration activates metasomatic activity. For example rising low-density *hypomagma* diapirs consume and contribute minerals to the rocks being intruded before solidifying as a batholith. **Magmatic/metasomatic cycles respectively formed the plutonic igneous and metamorphic crystalline bedrock basement in part from collapsed porous crust.**

Parentless Polonium Radiohalos

Fission decay of isotopes routinely progresses in nature. Let us reverse that process and consider isotope fusion in nature. Magma must harden into rock before it can retain the damage signature of radiohalos. Gentry (1986, 1988, & 1990) discovered that some polonium radiohalos in granite and pegmatite were parentless. The initial discolored spheres from the earliest decay steps were missing. However, adjacent radiohalos exhibited complete fission decay sequences.

The following decay series shows the sequence of fission steps proceeding from uranium-238 to lead-206.

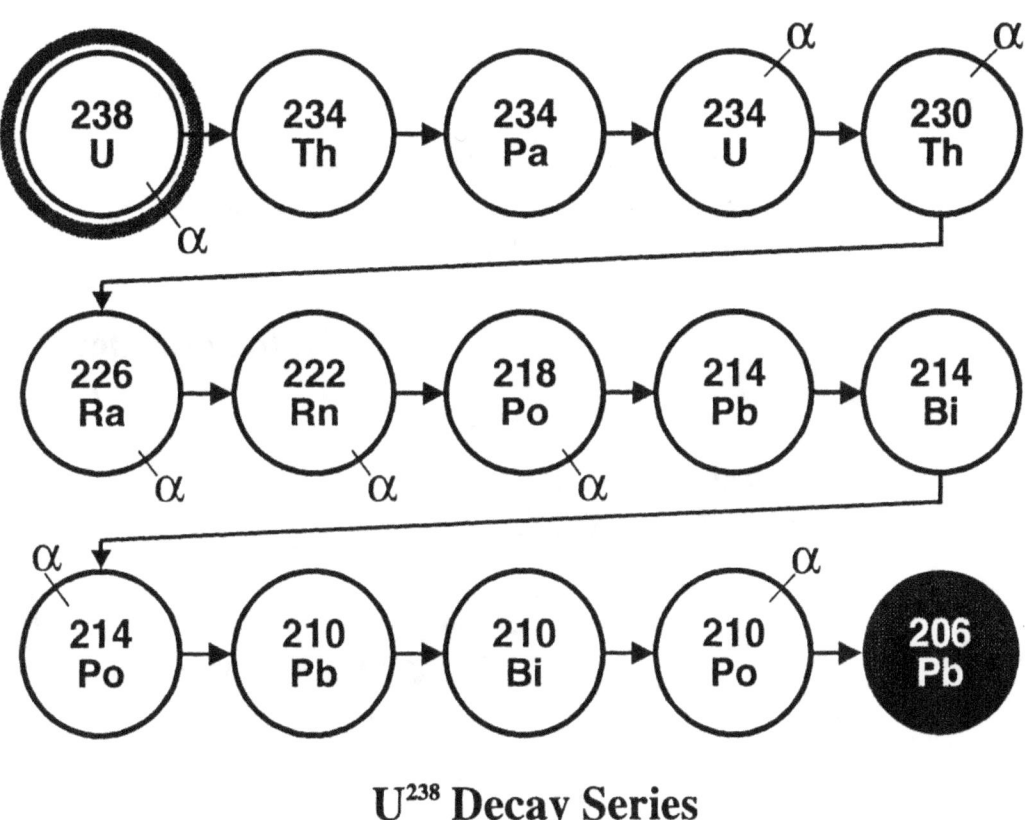

U^{238} Decay Series

α = Alpha Decay

Parentless polonium radiohalos are present in rocks exhibiting extensive geologic histories. Parentless polonium radiohalos have been discovered in dikes, cross-cutting older, host rocks (Wakefield and Wilkerson, 1990, pp. 329-344). Also, parentless polonium radiohalos are present in biotite from the Stone Mountain, Georgia (USA) granite pluton (Armitage, 2001, *TJ*, 15(1), pp. 86-88). *"The (Stone Mountain) granite intrudes both concordantly and discordantly into the country rock, which is composed primarily of biotite-plagioclase gneiss. The country rock was regionally metamorphosed to above the sillimanite isograde. At the granite contact, there is some evidence of contact metamorphism. Contact and structural data indicate that the granitic intrusion was late metamorphic and linked with the regional deformation associated with the uplift of the Southern Appalachians. The granite contains abundant xenoliths of the country rock"* (Walker, 2001, *TJ*, 15(1), p. 15). This extensive geologic history discredits the concept that parentless polonium radiohalos are exclusively characteristic of primordial Creation Week rock (Snelling and Armitage, 2003, pp. 255, 260).

RADIOHALO SPHERES OF DISCOLORATION
Diagram below from: *The Young Earth*
By: John Morris, 1994, p. 63.

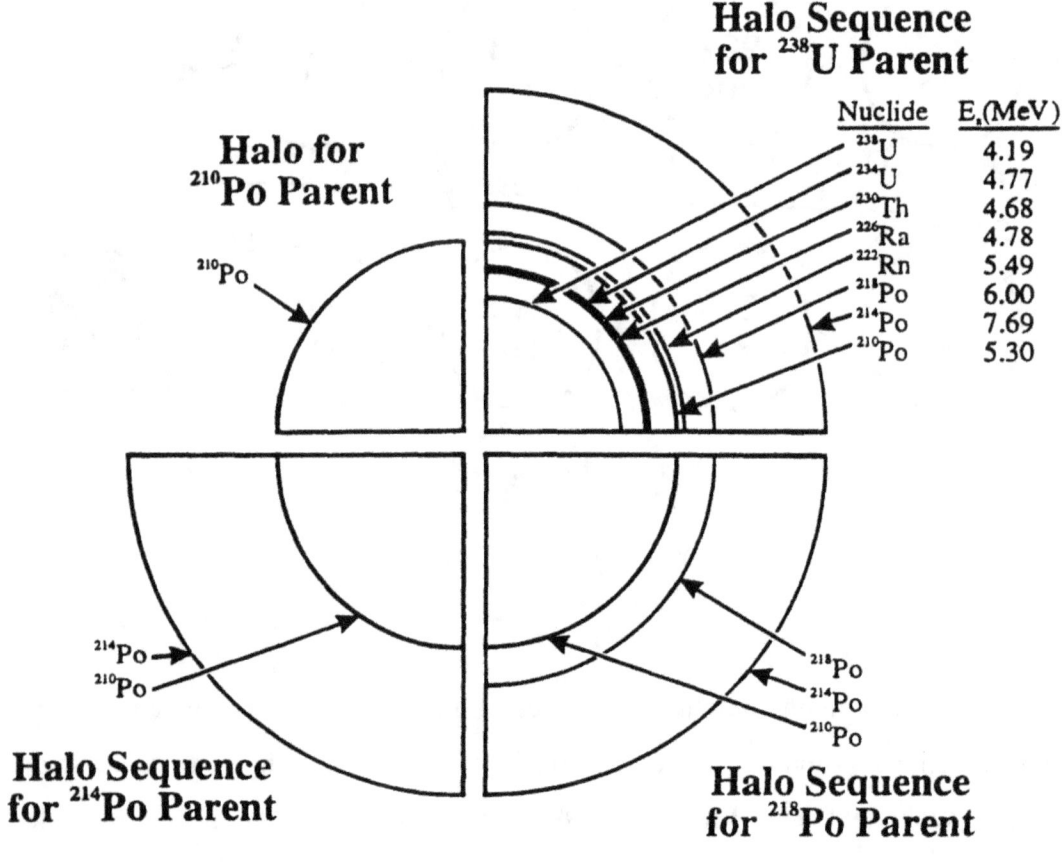

Characteristic ring configurations for different parent elements.

Alpha decay generates a specific diameter sphere of discoloration in the mineral matrix surrounding the radioactive parent isotope. The diameter of the damaged zone is a measure of the level of decay energy released for each alpha decay step in the series. Therefore, each radioactive decay series produces a specific pattern of concentric spheres of discoloration. Beta decay stages represent electron emission which does not decrease the isotope's atomic weight number.

Mineral solidification depends upon specific temperature, pressure, oxidation/reduction potential, pH, etc. conditions. Parentless polonium radiohalos have been discovered in metamorphic bedrock across the Great Smoky Mountains of Tennessee. The polonium (Po) radiohalos below were concentrated along the boundary between the Garnet and Staurolite Metamorphic Zones (Snelling, A.A., 2005, pp. i-iv).

Isotope	Total Radiohalos	Half-life
Po-218	2	3.1 minutes
Po-214	152	164 microseconds
Po-210	1857	138 days

(from Snelling & Armitage, 2003, Table 1)

A uranium-initiated fission progression would be expected to exhibit a uniform number of total radiohalos for each of the above polonium isotopes along the decay sequence in solid bedrock. The 138-day half-life for Po-210 might be expected to provide sufficient solidification time to preserve Po-210 radiohalos in greater numbers. However, the Po-214 half-life of 164 microseconds would be expected to preserve an essentially equal number of radiohalos as Po-218, but does not. This implies some other process is responsible for directly forming Po-218, Pb-214, Bi-214, Po-214, Pb-210, Bi-210, and Po-210, which form parentless polonium radiohalos from alpha decay energy emissions. The steady decrease in "Total Radiohalos" from Po-210 (1857) to Po-214 (152) to Po-218 (2) implies the process began with Po-210 and progressively declined in frequency in Po-214 and Po-218. **The implication is that Pb-206 was being fused into abundant Po-210, lesser amounts of Po-214, and a trivial amount of Po-218.**

Sonofusion

Sonofusion can theoretically be caused by any sudden shock compression generating sufficient activation energy to initiate the collision fusion process. Sonic pressure waves generate intermittent expansion and collapse of microscopic bubbles or liquid metal. Collapse of the bubbles (cavitation) produces extreme pressure which generates light (sonoluminescence). The conversion of sound to light is equivalent to an energy amplification of eleven orders of magnitude. Each bubble collapses to near its minimum radius, generating a very intense shock wave; hence, producing extreme temperatures and pressures within the bubbles or liquid metal.

The work of Adamenko artifically simulates a sonofusion-like mechanism. *"Ever since the beginning of the year 2000 numerous experiments were carried out at the Proton-21 Electrodynamics Research Laboratory in the Ukraine* (www.proton21.com.ua). *The results of these studies demonstrate that all known, stable chemical elements, plus a number of their isotopes are produced via energy explosion-induced compression of not only any single metallic element (copper, silver, lead, etc.) but even an organic substance such as polyethylene. Furthermore, not only were all the normal stable chemical elements produced, but there was even evidence that the unstable so called super-heavy elements were also produced; however, these were readily converted into other stable elements* (see Adamenko, et. al, 2007). *The most significant features of these results are two-fold: 1) all known chemical elements can be produced from another pure substance containing one or more pure elements; 2) the production of all other known elements occurs instantly upon explosion-induced compression of the target substance"* (Boudreaux and Baxter, 2010, p. 37).

Water provides an abundant source of oxygen nuclei for sonofusion with lead by expelling carbon as a fusion by-product. *"Lead has four stable isotopes: Pb-208 (52.3% abundance), Pb-207 (22.1% abundance), Pb-206 (24.19% abundance), and Pb-204 (1.4% abundance), so any of these isotopes could be involved in fusion with oxygen isotopes (O-16, O-17, O-18). For example: Pb-208/82 + O-16/8 fuses to Po-210/84 + C-14/6 + approx. 1.4 MeV (energy excess). The collision activation energies for all of these Pb/O processes producing Po are about the same order of magnitude (3MeV)"* (Boudreaux, 2010, pers. comm.).

The Antediluvian crust solidified with abundant volatile fluids encapsulated in tiny vugs within the solidified rock fabric. Post-Flood porous crust collapse generated "hammer-like" compressional events, producing sonic pressure waves that instigated sonofusion in bedrock. Lead's low temperature of solidification allows microscopic amounts of liquid lead to be encased within the matrix of most other solidified rocks across a widespread temperature/pressure domain.

Several kilometers below Earth's surface, atomic lead (as a molten trace component within solidified bedrock) was suddenly bombarded by sonic pressure waves from the collapsing crust above. The shock compression generated sufficient activation energy to trigger the sonofusion of lead and oxygen into unstable polonium isotopes, without melting the surrounding bedrock.

As soon as the collapse compression dissipated, pressures returned to the pre-collapse state. The fission decay sequence of the recently-generated polonium initiated alpha emissions, which damaged the surrounding solid rock, creating parentless polonium radiohalo signatures. Radiohalo signatures were recorded within the collapsing domain and beneath it. However, subsequent metamorphic alteration erased most radiohalo signatures within the collapsing domain. **Parentless polonium radiohalos are proposed to be the product of sonofusion generated by porous crust collapse.**

Continental Shields

"Continental shields" are relatively stable portions of Earth's crust composed of metamorphically-altered pre-continental crust. Continental shield areas formed during the Regressive Stage of the Worldwide Flood as ocean basin collapse generated underthrusting of adjacent continental crust in some regions. Associated tectonic disruption produced extensive fracturing of the porous pre-continental crust. Crust-forming magma below (*pyromagma, hypomagma* and *basalt*) was overpressured by tectonic compression and intruded into the fractured crust above. Hot intrusive volatiles and minerals comingled with and metamorphically altered the fractured crust above. Subsequent cooling and solidification formed tectonically stable continental shield areas.

Shield areas are characterized as Precambrian crystalline igneous and high-grade metamorphic rocks. Metamorphic rocks form by alterations of preexisting solid rocks induced by changes in temperature, pressure, shearing stress, and mineral supplementation-depletion-replacement associated with intruding magmatic fluids.

Intruding magmatic volatiles and low melting point minerals were also expelled into regressing floodwaters above via hydrothermal vents. Abundant low melting point minerals (predominantly iron and silica) spread outward from hydrothermal vents and settled to the bottom in colloidal form and/or by chemical precipitation.

Precambrian Banded Iron Formations

Banded iron formations (BIF) provide the world's largest iron ore deposits. The geologic process featuring colloidal settling and/or chemical precipitation to form large scale BIF is only characteristic of Precambrian stratigraphy. BIF is distinguished by thin (millimeter to centimeter scale) alternating layers of silica in the form of chert or jasper interbedded with iron minerals in the form of magnetite, hematite, or siderite.

Iron and silica interbedding was probably induced by rhythmic geochemical fluctuations during the precipitation process. *"We find that positive feedbacks occur among the chemical reactions when hydrothermal fluids mix with ambient sea water. These feedbacks lead to alternating precipitation of Fe and Si minerals, owing to the formation of complexes between Fe(II) and silicic acid ... We suggest that the small-scale (<1 cm) banding was produced by internal dynamics of the geochemical system, rather than any external forcing"* (www.astrobiology.nasa.gov/nai/..../alternative-formation-mechanisms-for-banded-iron-formations/; source: Wang, Y., et al., 2009, pp. 781-784). *"... both iron and silica may have been derived from submarine magmatic and hydrothermal activity ... Iron and silica-rich layers, originally deposited as **amorphous** gels, subsequently lithified to form banded iron formations"* (www.enotes.com/banded-iron-formations-reference/banded-iron-formations; source: Cengage, G., 2003, *World of Earth Sciences*).

The uniform appearance of banded iron formations from different shield areas on different continents around the world indicates that specific depositional process ceased

following Precambrian deposition. (In contrast, Phanerozoic ironstones appear to have been sourced by an entirely different mineral replacement process.)

ROCK RECORD CONTEXT

The Antediluvian surface was obliterated as floodwater inundated the land, carving the *Greatest Angular Unconformity*. The ravages of the Inundation Stage of the Worldwide Flood eroded, transported, pulverized, and deposited fine-grained sediments in areas where waters deepened and flow rates subsided. Only a few Antediluvian sedimentary remnants overly crystalline basement bedrock in topographic depressions (as evidenced by mud cracks). Tidal flow winnowing, erosional beveling, and sediment transport was primarily limited to shallow depths shifting upslope and downslope respectively in accordance with the initial inundation and subsequent regression of the Worldwide Flood.

During the Regressive Stage of the Worldwide Flood, ocean basin collapse generated continental underthrusting and horizontal compression of adjacent emerging continental crust. Resulting tectonic movement tilted unconsolidated "Proterozoic" Flood deposits. Meanshile, sheetflows retreating off the continents across the denuded land surface carved the *"Great Unconformity"* (not the earlier *Greatest Angular Unconformity)*. Uplifted blocks of "Proterozoic" sedimentary formations were truncated by the retreating sheetflow. The majority of continental shelf sediments were deposited during the Regressive Stage of the Worldwide Flood and for several years thereafter. However, as surface vegetation became established, post-Flood erosion, sediment transport, and deposition progressively diminished.

Ocean basin collapse provided space for floodwater retreat off emerging continents. However, ocean basin collapse displacement generated horizontal compression, underthrusting and uplifting of continental rims. Raised continental rims blocked floodwater drainage off continental interiors. Shallow continental interior seas remained "perched" above ocean levels at the end of the Worldwide Flood.

Continental basin sedimentary formations commonly exhibit an irregular "bowl-shape", with thick central basin deposits thinning to a feather edge on the continental basin periphery. Sedimentary formations typically exhibit deep water ecosystem facies near the basin center extending outward to shallow water facies on thinner basin flanks. That indicates continental basin subsidence occurred during deposition. The best explanation for continental basin subsidence is the progressive collapse of porous crust at depth. Continental basin seas were sustained as long as collapse volumetrically exceeded imported sediment and precipitate infilling.

During the Post-Flood Millennium, remnant *pyromagma* beneath the collapsing continental crust periodically escaped upward towards the surface and was extruded, as calcium carbonate-rich volatiles into shallow continental interior seas above. Extruding ash and precipitate killed marine life and settled onto the seafloor, providing optimum Paleozoic-style fossil-forming conditions.

The modern subsurface concentration of pyroclastic minerals has been severely depleted through time. A large percentage of available pyromagma volatiles have already escaped from Earth's interior into Earth's atmosphere and oceans and comingled with sedimentary deposits. Also, Earth's modern crust is too fractured to restrict rising volatile migration and encapsulate a high proportion of liquid-filled vugs in newly solidifying bedrock. Therefore, the conditions necessary to form highly-porous crust are no longer available. Consequently, porous crust that collapsed ocean basins during the Worldwide Flood and frequently collapsed porous crust on continents during the Post-Flood Millennium was not replaced. Also, the level of modern collapse is deeper and less impactive at the surface than shallower collapse events that occurred routinely during the Post-Flood Millennium. As a result, *pyromagma* eruptions have become increasingly rare in modern times. Also, earthquake activity declined exponentially after the final centuries of the Post-Flood Millennium. At that time, catastrophic mountain-building events of the Late Pliocene and Pleistocene accelerated earthquake activity.

Biological Evolution did not cause the Geologic Column fossil sequence. **Instead, the fossil record reveals a post-Flood ecosystem recolonization sequence.** However, that recolonization sequence progressed at different rates in each continental basin in accordance with changing conditions. Consequently, the following stratigraphic nomenclature correlates to ecosystem recolonization stages instead of geologic time.

EARTH'S CHANGING STRATIGRAPHY

GEOLOGIC COLUMN	COLLAPSE TECTONICS	DISTINGUISHING CONTINENTAL LITHOLOGIC CHARACTERISTICS
Holocene	Modern Day	Modern depositional regimes Few terrestrial body fossils
Pleistocene	Ice Age	Glacial moraines and bedrock striations Mountain range upheaval
Tertiary	Late Post-Flood Millennium	Subaerially-deposited clastic materials Decreasing terrestrial body fossils
Mesozoic	Middle Post-Flood Millennium	Lowland and nearshore clastic deposits Abundant terrestrial body fossils
Paleozoic	Early Post-Flood Millennium	Shallow inland sea carbonate precipitates Abundant marine body fossils
Proterozoic	Post-Flood Ediacaran Recollonization	Mudcracks, soft corals, jellyfish species cyanobacteria species (no body fossils)
	Worldwide Flood Deposits	Fine sand and smaller grain sizes Indistinct formation boundaries Undifferentiated organic materials Fluid evulsion structures (water expulsion)
	Antediluvian Age	Pre-Flood mudcracks in basal formations
Archean	Worldwide Flood Subsurface	Greenstone belts and granite-gneiss complexes (intursives comingled with porous crust)

PROTEROZOIC WORLDWIDE FLOOD

Proterozoic sedimentary formations provide the optimum stratigraphic evidences correlating to cataclysmic Worldwide Flood deposits. Exceptions include 1.) mudcracks in basal Proterozoic deposits corresponding to pre-Flood sediments and 2.) uppermost Proterozoic Ediacaran strata corresponding to very early post-Flood recolonization. However, the other Proterozoic sedimentary formations correspond to Worldwide Flood deposits.

Shallow, rapid surface water flows facilitate erosion. Proterozoic sedimentary formations are characteristically bracketed below and above by the two most extensive unconformities on Earth. The transgressional *Greatest Angular Unconformity* at the base of the Proterozoic was eroded by intense rainfall and surface runoff during the initial 40 day Worldwide Flood Inundation. The regressional *Great Unconformity* at the top of the Proterozoic was initially eroded by a rapid sheetflow retreat of floodwater off emerging continents during the Worldwide Flood Regression.

Worldwide Flood deposits exhibit no lengthy brakes in time between erosion and deposition. Sedimentary formation boundaries within the Proterozoic are characteristically less distinctive than Phanerozoic formation boundaries. That implies relatively continuous deposition transitioning from one Proterozoic sedimentary formation to the next during the Worldwide Flood Year.

Worldwide Flood catastrophism correlates to the highest-energy erosion and transport mechanisms of the past, resulting in pulverization of Flood deposits into fine-grained materials. Proterozoic sedimentary formations comprise the largest collection of fine-grained materials the Rock Record. For example, *"... the existence of black shale, a shale containing 3 to 15% organic carbon that is especially common in the Precambrian and Paleozoic (see Pettijohn, 1975, pp. 282-284). Black shale is difficult to explain within secular geology. Its existence within both Precambrian and lower Phanerozoic sedimentary rocks would indicate it formed during the Flood, and that the pre-Flood/Flood boundary is low within the Precambrian"* (Oard, 2013b, p. 7).

"Another unique type of rock is quartz arenite, a type of sandstone characterized by greater than 95% quartz that is very well rounded and highly sorted (see Pettijohn, 1975, pp.

230-235). *Quartz arenite is commonly found in the late Precambrian and early Paleozoic. It can sometimes be quite thick, up to 1000 m, and extensive, such as the Athabaska Formation of northern Saskatchewan, Canada, which covers about 104,000 km^2 and the Thelon Formation of similar extent in the Northwest Territories of Canada ... It probably represents winnowed sand in a highly turbulent environment with a large amount of sediment in the water. The fact that quartz arenite transcends the Precambrian/Cambrian boundary indicates that this boundary is not the pre-Flood/Flood boundary in sedimentary rocks"* (Oard, 2013b, p.7). Collapse Tectonics proposes extensive quartz arenite deposits correspond to materials catastrophically-eroded during the Worldwide Flood and subsequently transported and deposited during the Late Flood Regression Stage and continuing into early post-Flood time. **In conclusion, Proterozoic sedimentary formations provide the best stratigraphic evicence corresponding to Worldwide Flood deposits.**

POST-FLOOD MILLENNIUM

The Post-Flood Millennium corresponds to a tectonically-active period on Earth's continents following the Worldwide Flood. **Post-Flood Millennium stratigraphy begins with Late Proterozoic Ediacaran ecosystem recolonization and culminates with Late Pliocene and Pleistocene mountain-building.** The thousand-year "millennium" designation provides a rough estimate of the Collapse Tectonics time-frame allotted to the following sequence of geologic evidences.

Worldwide Flood erosion had beveled topographic highs and infilled topographic depressions with sediments. The beginning of the Post-Flood Millennium was flatter than the Antediluvian land and lakes topography. Shallow continental seas and vast denuded playa areas covered continental interiors. However, tectonic activity had transformed continental surfaces into a topographic configuration similar to modern continents by the end of the Post-Flood Millennium.

Stromatolite Recolonization

Paleo-topographic depressions on emerging continental surfaces acted as sediment depositories for suspended materials in retreating floodwater. Intermittent collapse beneath fracture-breached crust periodically expelled hot mineral-rich brines into shallow continental interior seas above, contributing calcium carbonate, mineral ash, and precipitates to accumulating seafloor deposits. Surface muds washed into depressions from the surrounding subaerial landscape, mixing with brine components. These typically thinly-layered deposits featured stromatolites and ripple marks in shallows, winnowed beach sands on flanks, and mud cracks in subaerially exposed areas.

"Proterozoic" stromatolites correspond to very early post-Flood recolonization in shallow marine environments. Stromatolites are reef-like remnants thought to be precipitated mineral matter on microbal communities, primarily blue-green algae, growing by photosynthesis" (Ferrell, 2005, p. 444). *"Although stromatolites in Precambrian sediments are rather common, often large and extensive, and occasionally huge, they are rare in Phanerozoic sediments and both rare and small in the present world ... It is thought that grazers prevent organic mats from getting large enough to form large stromatolites ... Optimum conditions for stromatolite development would seem to be 1) intertidal to shallow water for optimum photosynthetic activity and thus optimum organic mat production; 2) high precipitate production to maximize accretion rate of inorganics; and 3) a lack of grazers. A shallow water intertidal hydrothermal regime would meet all three conditions"* (Wise, 2003, p. 360). Early post-Flood conditions allowed stromatolite development to briefly flourish as extruded volatile brines poisoned grazers. However, as volatile brine expulsion subsided and waters became more hospitable, grazers repopulated and restrained stromatolite growth.

By the uniformitarian timescale, limestones and dolomites *"... are especially abundant in the Proterozoic (between 2.5 billion and 542 million years) and the early Paleozoic"* (Oard, 2013b, p. 7). Collapse Tectonics correlates these carbonate deposits to porous Antediluvian crust that extruded volatiles during intermittent collapse events on post-Flood continents.

Worldwide Flood tectonics caused perched continental interior seas to be hot and highly-mineralized due to magmatic contributions and high heat flow through the disrupted crust. Lower-density fresh water spread out on top of denser brines in closed continental basins. **Marine ecosystems developed according to depth and salt tolerance of inhabiting species.** Outlet rivers drained near-surface fresh water from continental interior seas into oceans. However, as outlet rivers eroded deeper and sediments filled basins, brine-rich waters increasingly overflowed basin rims and progressively dispersed into oceans. Fresh water ecosystems became more prevalent as the Post-Flood Millennium ran its course.

Ediacaran Recolonization

The first post-Flood multicellular fossils in the Rock Record are referred to as "Ediacaran fauna". *"Ediacara-type faunas are now known on all continents except Antarctica"* (Monroe and Wicander, 1994, p. 550). The Geologic Column designates these soft-bodied organisms as Late Proterozoic in age. Ediacaran animals probably lived in a shallow, continental interior sea ecosystem.

Rocks in the Ediacara Hills of southwestern Australia contain imprints of soft-bodied jellyfish, segmented worms, and soft corals. *"... These organisms were pioneers that could cope with conditions that would have killed other animals. They thrived for a time, but soon gave way to the main wave of recolonization"* (Tyler, 2006, p. 79). *"Since many deep ocean creatures are large, flat, soft-bodied organisms such as the Ediacaran organisms are, perhaps the Ediacaran biota inhabited the deepest waters of this epicontinental lagoon"* (Wise, 2003, p. 362).

Lower-energy post-Flood erosion generated Phanerozoic clastic sediments characterized by larger grain sizes than higher-energy Worldwide Flood erosion and transport processes. "Upper Proterozoic" sands are typically characterized as non-marine sands which grade upward into marine sands with increasing glauconite "greensands" higher in the formation. Glauconite is a marine mineral which is considered too fragile to survive transport. *"The excepted model has glauconite forming from the alteration of fecal pellets"* (Reed, 2005, p. 53). **Therefore, "Upper Proterozoic" sands are herein identified as early post-Flood deposits.** Fecal pellets initially provided a miniscule contribution to rapidly-deposited "Upper Proterozoic" glauconite sands. However, slowing sedimentation rates and recolonization produced an upward increase in fecal pellet concentrations.

Cambrian Recolonization

"Cambrian rocks, often with an unconformity at their base, are of worldwide occurrence" (Robinson, S.J., 1996, p. 39). Cambrian marine sediments above the *Great Unconformity* exhibit an astounding abundance of plant and animal fossils. *"The "explosion" restored the next section of the food chain, including mollusks, brachiopods, sponges, jellyfish and arthropods (including trilobites)"* (Tyler, 2006, p. 80). Plants, including spores from advanced flowering plants, and at least one representative of every animal phylum have been found in Cambrian strata (Ferrell, 2005, p. 442).

The characteristic abundance of trilobite fossils in the "Cambrian" ecosystem is attributed to their high tolerance for murky, brine-rich, post-Flood seas. As waters progressively became clearer and fresher, predators flourished, reducing and eventually exterminating most trilobite species.

The base level of the food chain was established in each area before the next level of the food chain could invade, inhabit, and proliferate. Species unsuited to most post-Flood environments temporarily had to survive in isolated locations where conditions were tolerable. Minor variations in the chemical components of continental interior seas had a major impact upon which organisms proliferated at the base of the food chain. Successful organisms supplied the food chain succession that fed upon them. This proliferation of Cambrian fossils had nothing to do with a sudden acceleration of biological evolution. **Instead, the Cambrian Explosion corresponds to early post-Flood recolonization, facilitated by rainwater dilution of caustic brine waters, and optimum fossil preservation conditions in shallow continental interior seas.**

Paleozoic Carbonate Deposition

The massive volume of calcium carbonate stored in Paleozoic limestones and dolomites in continental basins begs for a source. Some Paleozoic formations exhibit abundant shell debris approaching the outside edge of their formation boundaries. However, Paleozoic carbonates appear to have been sourced by subterranean extrusives.

Geologists have often puzzled over the reason why ancient limestones are so compositionally different, from the slowly-accumulating, shell material lime muds observed in modern deposits. Rock matrix texture and the common presence of cavaties (void spaces) discredit shell accumulation as a major depositional mechanism for extensive Paleozoic limestone formations in continental basins.

Marine organisms slowly absorb calcium carbonate from seawater to build their shells. Upon death, the low solubility of calcium carbonate prevents these shells from being dissolved back into seawater. Modern lime muds form by the breakdown of skeletons of carbonate-containing marine organisms (Austin, 1990, No. 210, p.i). The product is silt-sized crystals approximately 20 microns in diameter. These lime muds are composed of 60% to 95% aragonite and 0% to 10% calcite (Steinen, 1978, p. 1140).

In contrast, ancient lime mudstone (micritic limestone) is composed of clay-sized crystals less than 4 microns in diameter. Micritic limestone is composed of nearly 100% calcite or dolomite matrix with larger shell fragments floating in that fine crystal matrix (McKee and Gutschick, 1969, p. 103).

Modern lime muds and ancient micritic limestones also differ in crystal shape, size, and distribution. These differing textures are frequently attributed to recrystallization during de-watering and consolidation into limestone. However, recrystallization generates larger crystals from smaller ones. In contrast, conversion of lime muds into micritic limestone would require a reduction of crystal size. Therefore, ancient limestones were not recrystallized from accumulated shell material.

Early researchers on the microcrystalline calcite (micrite) ooze of ancient limestone argued that it formed by direct precipitation from seawater (Folk, 1959, p. 8). A non-biogenic source for the bulk of ancient limestones is still supported by the evidence. It is reasonable to propose that the abundant supply of calcium, magnesium, and carbon dioxide required to form Paleozoic carbonates was sourced by submarine, calcite-rich, *pyroclastic* eruptions. In support of a subterranean source, *"Carbonatite is a volcanic rock consisting largely of igneous calcite, and suggests vast accumulations of carbonate at the base of the crust"* (Ollier and Pain, 2000, p. 180). Petrology literature contains about 330 descriptions of carbonatite-bearing volcanoes and intrusive complexes (Woolley, 1989, p. 15).

FOSSILIFEROUS "PALEOZOIC" LIME MUD DEPOSITION
IN POST-GENESIS FLOOD CONTINENTAL BASIN SEA

PRE-ERUPTION CONDITION

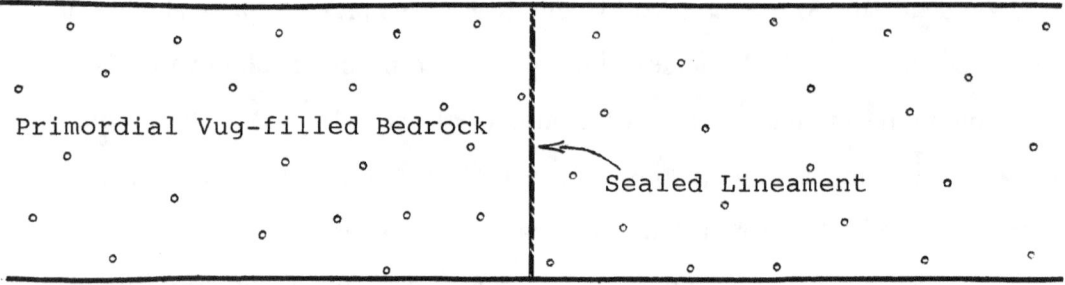

SUBMARINE ERUPTION AND LIME MUD DEPOSITION

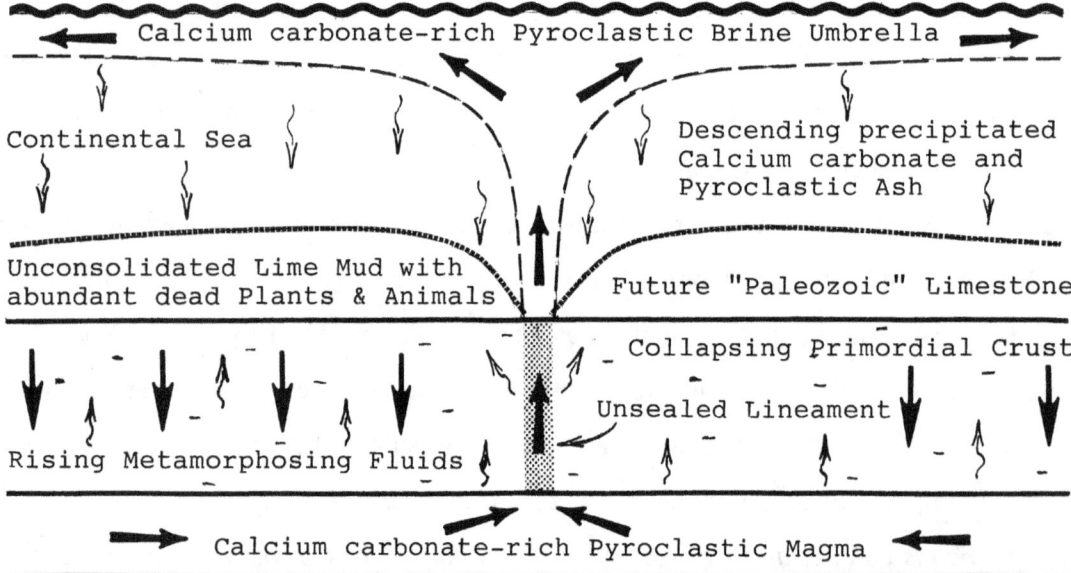

During the Post-Flood Millennium, porous continental crust periodically collapsed below continental interior seas, releasing calcium carbonate-rich volatiles. Lineament fissure seals ruptured, opening fracture conduits upward to the seafloor. Scalding, calcium carbonate-rich volatiles erupted into the sea water above. The scalding *pyroclastic* brine rose from the erupting vent to the surface and spread outward across the shallow sea. *Pyroclastic* gases bubbled to the surface and spread outward, hovering above the sea in a poisonous dense cloud. The *pyroclastic* cloud shed hydrogen sulfide and ash into the continental basin sea. Hydrogen sulfide within the *pyroclastic* brine absorbed oxygen from the seawater to form sulfuric acid. Sulfuric acid descended toward the sea floor, rupturing gills and killing sea life.

Calcium carbonate ash also "rained" down upon unsuspecting sea life below. Panicked fish frantically swam this way and that, as they desperately attempted to escape the descending toxic brine. Local marine ecosystems were entombed and preserved in lime mud by acidic, reducing conditions. Acidic conditions mobilized minerals and accelerated the hardening of lime mud into limestone.

Paleozoic limestones are frequently characterized by abundant cavaties within the rock matrix. The cavaties range from millimeters to meters in diameter and sometimes constitute 20% or more of total rock volume. Limestone voids are explained as sediment degassing features. *"Acidifying the solution increases the* (calcium carbonate) *solubility, because the carbonate ions are chemically hydrolyzed, forming unstable carbonic acid which decomposes into carbon dioxide and water"* (Boudreaux, 2006, pers. comm.). Lime mud solidified rapidly into limestone, encasing degassing cavities within the rock matrix. Numerous and larger-sized cavities, were preserved in limestones located close to erupting *pyroclastic* vents.

The solubility of a metal salt in seawater *"depends greatly on the lattice energy of the crystalline salt and the ability of the solvent dipoles to overcome the lattice energy and separate the ions in solution ... Generally speaking, when the lattice energy reaches some 2600-2700 kcal/mole, the water dipoles are not able to overcome this energy ... Calcium carbonate has a lattice energy of 2800 kcal/mole and is thus not soluble in water to any measurable extent (0.0007 grams per 100 grams of water), at any temperature"* (Boudreaux, 2006, pers. comm.).

In contrast, other mineral salts which are precipitated as evaporates are readily soluble and carried in much higher concentrations in seawater. Evaporite precipitation from seawater should make associated calcium carbonate precipitation insignificant by comparison. However, Paleozoic evaporates are volumetrically "overwhelmed" by the abundant volume of precipitated carbonates. Therefore, evaporation is eliminated as a significant depositional mechanism for Paleozoic limestones. **The most logical source of Paleozoic limestones and dolomites is direct intrusion of carbonate-rich volatiles escaping from collapsing porous crust into the seawater above.**

Entombed Fish

The Old Red Sandstone of northern Scotland contains abundant fossil fish preserved in a terrified attempt to survive. Preservation of complete skeletons indicates the fish were buried before scavengers could destroy the skeletons. The fossil fish *"are contorted, contracted, curved; the tail in many instances is bent around to the head; the spines stick out; the fins are spread to the full, as in fish that die in convulsions"* (Velikovsky, 1955, pp. 29-30). The area of annihilation covers perhaps ten thousand square miles (Miller, H., 1865, p. 222).

Fossil fish from Monte Bolca, near Verona, in northern Italy, perished suddenly as they were entombed in limestone and calcareous slate. These complete fish skeletons are densely packed together parallel to the laminations of the strata. Some fish are preserved with traces of color upon their scales, indicating they were entombed before their soft parts decomposed (Buckland, 1937, p. 101). The fish were poisoned, cooked, or suffocated by lime mud clogging their gills before sinking to the bottom and being immediately blanketed by calcareous sediments.

Fossil fish writhing in the agonies of death are common worldwide. The variety of rocks in which these entombed fish are preserved, depends upon ash and chemicals volcanically extruded into seawater immediately prior to entombment. Famous examples include the cupriferous slate of the Harz Mountains of Germany; the bituminous slate of Mansfield in Thuringia; the coal of Saarbrucken on the Saar; the calcareous slate of Solenhofen; the blue slate of Glaris; the marlstone of Oensingen, Switzerland; the black limestone of Ohio and Michigan; the Green River Formation of Arizona; and the diatom beds of Lompoc, California (Velikovsky, 1955, p. 31).

Depositional variables often dictated which organisms were preserved in fossil form. For example, sulfur-rich "Devonian" deposits commonly contain abundant fish fossils. **Devonian formations correspond to a killing zone rather than an "Age of Fishes".** Their position in the Stratigraphic Record implies volatile expulsions achieved maximum hydrogen sulfide concentration at that stage in each continental basin. Lower hydrogen sulfide concentrations in sediments above and below the "Devonian" deposits allowed fast-swimming fish to escape the carnage while slower marine life was killed, buried, and fossilized in the habitat where they had lived.

Fossil Reefs

True fringing reefs and atoll reefs commonly observed today are rare in the geologic record. In contrast, fossil reefs are commonly patch reefs or true barrier reefs that separate open marine deposits from lagoonal facies (Selley, 1970, p. 158).

Post-Flood porous crust collapse tilted lineament-bounded blocks on the seafloor above. Upward-tilted corners close to the surface accommodated patch reef formation, or a variation called pinnacle reefs. Barrier reefs formed where a continuing series of lineament-bounded blocks were tilted side by side, cresting at shallow depth.

Porous crust collapse events ruptured lineament seals adjacent to buried reefs. Calcium carbonate-rich volatiles were expelled, accelerating growth of reef-building organisms. Reefs are also frequent hosts to metalliferous deposits. Mineralized magmatic fluids migrated up the opened lineament fissures into the porous reef material. Lower Carboniferous reefs of Ireland have mineralized cores (Derry, et al., 1965, pp. 1218-1237). Devonian reefs of the Canning Basin, Australia, are mineralized in the fore-reef facies (Johnstone, et al., 1967, pp. 599-612). Mineralization occured in lagoonal back-reef facies in Alpine Triassic reefs (Selley, 1970, p. 179). **The high frequency and intensity of collapse tectonic events during the Post-Flood Millennium induced compression-driven intrusion of calcium carbonate resulting in rapid reef growth, in some cases accompanied by mineralized magmatic fluids.**

Ecosystem Reversals

Continental basin collapse and sediment infilling proceeded from brine-rich, deep water ecosystems at the base upward to progressively shallower, fresh water ecosystems. However, collapse events also shifted from one portion of a continental basin to another as time proceeded. *"The sedimentary fill in parts of the Williston Basin area shows strong variations, both laterally and vertically."* These fill characteristics *"offer the best evidence of the changing patterns of tectonic activity through time, and which strongly imply the existence of discontinuous centers of activity"* (Krumbein & Sloss, 1963, p. 417). Correspondingly, ecosystems where shifting back and forth, adjusting to the specific location of the most recent continental basin collapse event. **Ecosystem reversals occurred when intermittent tectonic activity temporarily caused physical conditions in an area to revert to an earlier stage of the recolonization sequence.** So-called "overthrust belts" exhibit supposedly older index fossils overlying supposedly younger index fossils. These are examples of ecologic reversals to a characteristic earlier Post-Flood recolonization stage.

Hexapoda Gap

The Hexapoda Gap mystery questions why terrestrial insects are so abundantly fossilized in Carboniferous (Mississippian and Pennsylvanian) strata, but almost entirely absent in underlying Devonian formations. During the Post-Flood Millennium, terrestrial insects proliferated in areas surrounding shallow continental basin seas. Carboniferous depositional environments included swamps and shallow nearshore locations where insects proliferated and were routinely buried. In contrast, Devonian stratigraphy corresponds to a deeper water marine environment where intermittent porous crust collapse events were concentrated. Flying insects blown out to sea were typically consumed by fish and other marine inhabitants before burial and fossilization occurred. As continental basin seas shrank in size, Devonian ecosystems retreated toward basin centers. Carboniferous depositional environments also migrated toward continental basin centers, overlapping Devonian formations. **The sudden upward proliferation of terrestrial insect fossils in**

Carboniferous formations is best explained by a simple regression of continental basin seas – not sudden worldwide insect evolution and proliferation.

Evaporite Precipitation

Unlike calcium carbonate, most other common mineral salts are readily soluble and precipitate when specific mineral concentrations exceed the capacity of water dipoles to hold them in solution. Evaporation of seawater increases remnant mineral concentrations, causing those minerals to sequentially precipitate as "evaporites".

Water dipoles are normally able *"to overcome the lattice energy and separate the ions in solution"* of mineral salts with a lattice energy lower than the general 2600-2700 kcal/mole threshold. *"Sodium and potassium carbonates, for example, have lattice energies below 2500 kcal/mole: hence, they are readily soluble in water. Their solubilities decrease with lower temperature"* (Boudreaux, 2006, pers. comm.).

"Salvation energy" is either released or absorbed as a mineral salt precipitates or dissolves. Salvation energy is released as sodium and potassium carbonate salts precipitate. Therefore, precipitation of sodium and potassium carbonates is facilitated by seawater cooling. In contrast, *"Lithium carbonate has a lattice energy of some 2500 kcal/mole and exhibits a low solubility"* (Boudreaux, 2006 pers. comm.). So salvation energy is absorbed during precipitation. Therefore, heating of seawater facilitates lithium carbonate precipitatation.

Other precipitate minerals (including gypsum and halite) were also expelled during volcanic extrusions. In modern times, shallow continental sea bottoms leak hot magmatic fluids as black smokers along lineament ruptures. Halite and gypsum precipitate where black smokers expel mineralized brine in isolated shallows. In conclusion, each mineral salt has a unique regime of conditions which instigate precipitation. **Mineral salts precipitated in sequence according to changing physical conditions.**

Post-Flood Terrestrial Climates

Earth's early post-Flood atmosphere was warmed by heat transfer through the disrupted crust. Evaporation proliferated as hot oceans and continental interior seas steamed into the cooler atmosphere above. Ocean basin collapse released a large volume of carbon dioxide into the atmosphere, contributing to a post-Flood "green house" warming effect. Abundant atmospheric carbon dioxide generated prolific plant growth. Terrestrial ecosystems were temporarily established in irregular concentric rings around shallow continental interior seas in accordance with topographic position. During the Post-Flood Millennium, burial of a large volume of biologic materials in subsiding continental basins slowly reduced atmospheric carbon dioxide to modern levels.

Following the Worldwide Flood, differential sunlight heating between continents and oceans generated modern large-scale weather patterns. Those distinctive weather patterns diversified climates and caused a wide variety of unique ecosystems. Post-Flood temperatures fluctuated more and climates were more hostile than Antediluvian conditions. Descendants of Ark animals migrated into ecosystem niches where they could compete and survive. Terrestrial fossils became more frequently preserved higher in the Phanerozoic Rock Record as Ark descendents proliferated and spread.

Animal kinds migrated into a variety of post-Flood habitats. Members of each animal kind best suited to surviving to reproduction age passed on "favored" genetic information to their offspring. Elimination of descendants unsuited to each ecosystem resulted in a loss of disadvantageous genetic information. Separate regional populations diversified from each other according to lost genetic information. Initial success by invading animals was rarely sustained. Different plants and animals gained a competitive advantage as ecosystem variables changed rapidly during the Post-Flood Millennium. Specialized species in unique ecosystems became extinct as the ecosystem expired and bedcame unavailable for future fossilization. Fossil preservation of animals descended from Ark voyagers, depended upon migration routes, changing ecosystem conditions, competition from other animals, and natural catastrophies. **Modern species are descended from more genetically diverse animal kinds.** Air-breathing terrestrial forms are descendents of Ark voyagers.

Mesozoic Ecosystems

For several centuries following the Worldwide Flood, the climate adjacent to continental basin seas was hot and humid. Swampy lowlands surrounding post-Flood continental interior seas provided an ideal ecosystem for dinosaurs to thrive. Their scaly skin repelled attacking insect swarms. Swamps helped support a large Sauropod dinosaur's weight as it lazily grazed on prolific vegetation and grew to massive size. Large tetrapod dinosaurs feasted primarily on carrion of the dead. Warm, moist soils provided optimum conditions for dinosaur egg hatcheries.

Vegetation was plentiful and the hot muggy climate and swarming insects drove off humans and other competing lifeforms. To avoid the oppressive high-humidity heat and insect infestation, warm-blooded mammals inhabited highland and ocean coast ecosystems, far from continental interior seas.

The top-down collapse of porous crust underlying post-Flood continental basin seas was initially characterized by frequent small collapse events. However, as sediments infilled collapsing basins, sediment overburden pressures became greater and the volume of porous crust collapsed during collapse events increased. **Those larger volume collapse events generated less frequent but progressively larger tsunamis inundating subaerial "Mesozoic ecosystems".**

An increasing intensity of tectonic processes is revealed by the sequential preservation of Mesozoic fossils along the flanks of shrinking continental interior seas. *"Dinosaur tracks are quite diverse in Triassic and lower Jurassic strata, but the greatest diversity of their body fossils is in the Cretaceous ... Fossil tracks of small dinosaurs and other reptiles are almost entirely absent in upper Jurassic and higher strata, but abundant large dinosaur tracks are found in Cretaceous strata ... After deposition of the Cretaceous strata, even the large dinosaurs and their tracks are no longer found in subsequent strata* (Snelling, 2009, Vol. 2, p. 747).

The Mesozoic (Triassic-Jurassic-Cretaceous) sequence of smaller dinosaur evidences at the base, transioning upward to progressively larger-sized dinosaur evidences at the top was generated by a progressive increase in sediment transport. Triassic topographic surfaces sloped gently down toward continental interior seas. Sediment transport in tsunami backwash was low velocity and extended across broad plains. Backwash sediment layers were typically thin and inadequate for burying larger dinosaurs. In contrast, Cretaceous tectonics were uplifting terrestrial areas adjacent to continental basin seas. Tsunami inundations became shorter and slope grades steeper. Tsunami backwash returned to continental seas more rapidly, carrying a much larger sediment load. Thick deposits accumulated at the edge of the continental seas where flow velocities suddenly decreased. These thick continental sea edge deposits were capable of blanketing large dinosaurs killed by the tsunami. Dinosaur footprints are common because tsunami survivors walked across the shoreline muds to feast on an abundance of dislodged vegetation and corpses.

Mesozoic Clastic Deposition

Continental rim upheaval cradled shallow seas, which blanketed large areas of continental interiors following the Worldwide Flood. During the initial post-Flood centuries, adjacent terrestrial areas were too flat to generate significant erosion and sediment transport into an adjacent continental basin. Continental basins sank as porous crust progressively collapsed beneath them.

However, porous crust collapse compression could not depress higher-density materials below. As porous crust collapsed deeper and deeper, downward compression was redirected laterally. Lateral compression shifted crust at depth outward and upward toward continental basin flanks, uplifting adjacent areas. Increasing topographic relief accelerated subaerial erosion, sediment transport, and deposition of clastic materials into adjacent continental basins. Clastic sediment contributions from adjacent terrestrial areas eventually overwhelmed carbonate contributions from escaping volatiles. That Paleozoic carbonate to Mesozoic clastic transition was correspondingly accompanied by increasing subaerial ecosystem evidences.

Although the Paleozoic to Mesozoic depositional transition followed the same sequence worldwide, each continental basin progressed at varying rates and concluded at different times. Therefore, the similar stratigraphic sequence of fossil ecosystems worldwide is a function of that ordered continental basin porous crust collapse progression.

SEDIMENT INFILLING OF CONTINENTAL BASINS

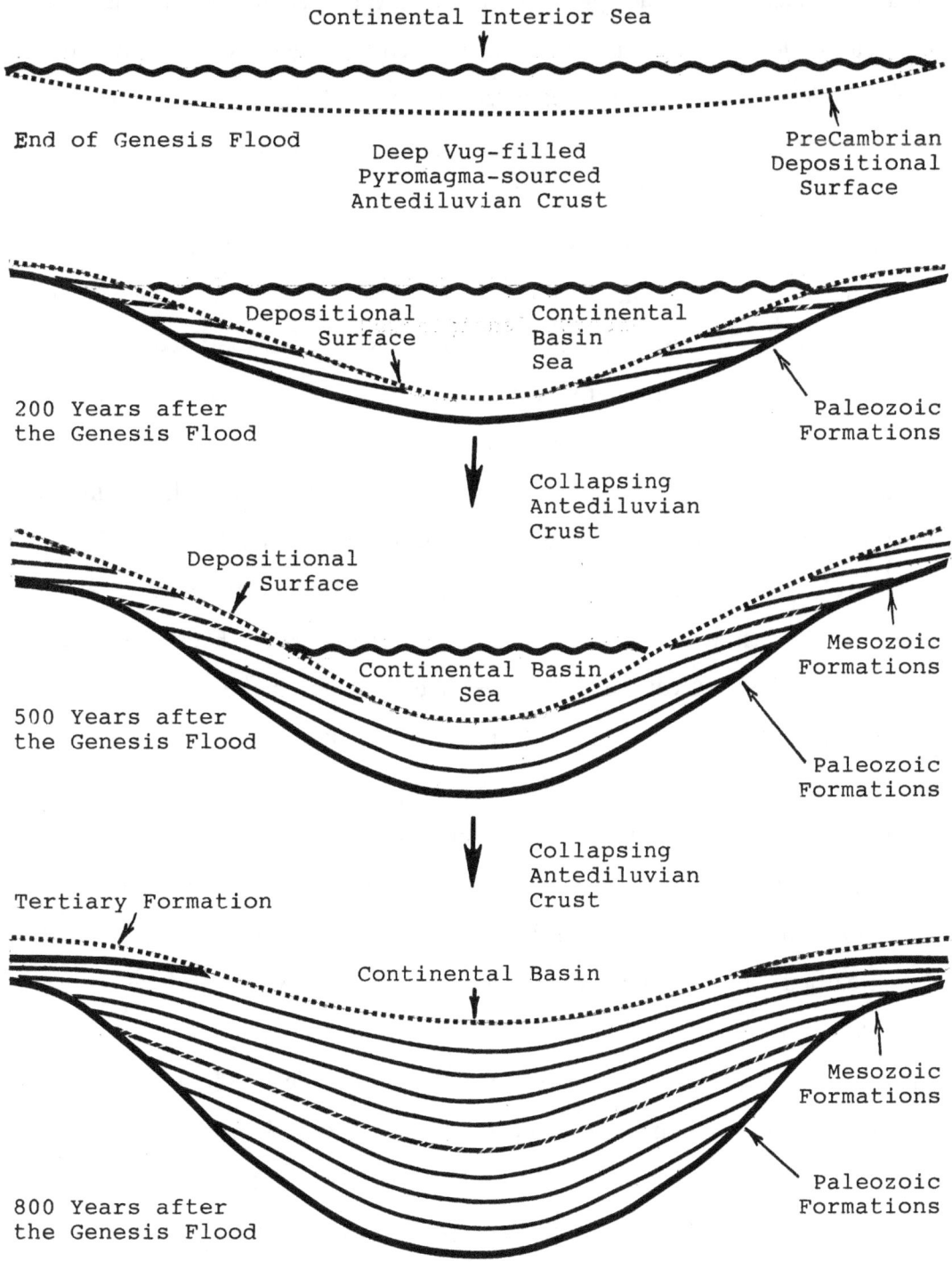

125

Tsunami Wave Backwash Deposition

Collapsing lenses of porous pyroclastic crust beneath a shallow continental basin sea caused the seafloor to drop, generating tremendous energy. Water rushed into the collapsing void from all sides, welled up, and rebounded outward. As the spreading wave entered shallower water, energy focused vertically, raising a massive tsunami wave.

The tsunami wave carried sea floor sediments and lime mud in suspension. That provided the raw materials for inland deposition of thinly-bedded carbonate layers, wrongly identified as marine deposits. Tsunamis initially extended inland for vast distances due to the low topographic relief. Those waters returned too slowly to transport sediments for long distances. Much of the tsunami water slowly returned to the continental basin sea below the surface as groundwater flow. Lifeforms were stranded on the surface to decompose.

As uplift progressively increased later during the Post-Flood Millennium, tsunami waves roared inland for shorter distances across adjacent swamps and lowlands. Ecosystems were successively ripped out by the onrushing tsunami deluge. The tsunami wave gradually lost momentum - then retreated back toward the inland sea. Ecosystem sediments, vegetation, and animals were carried back in the watery grasp of the retreating backwash and dumped as nearshore deposits, in the same relative topographic position as the materials were ripped out.

Depositional mixing of adjacent ecosystems was restricted by the short distance of inland transport and seaward retreat. **Higher elevation ecosystem deposits overlapped lower elevation ecosystems in "shingle-like" fashion.** (A noteworthy exception is floating plants being carried into and buried within the adjacent shallow marine exosystem by the tsumami backwash.)

The characteristic "inter-tonguing" of formations indicates both formations were deposited together at the same time. Each collapse event changed the shape of the continental basin sea and its surrounding ecosystem positions. Terrestrial sediments filled the continental basin faster than the continental basin sank. During the years and decades between intermittent tsunami events, ecosystems quickly reestablished prolific terrestrial vegetation adjacent to continental basin seas preceeding the next impending tsunami event.

CONTINENTAL BASIN SEA COLLAPSE LAUNCHING TSUNAMI WAVES

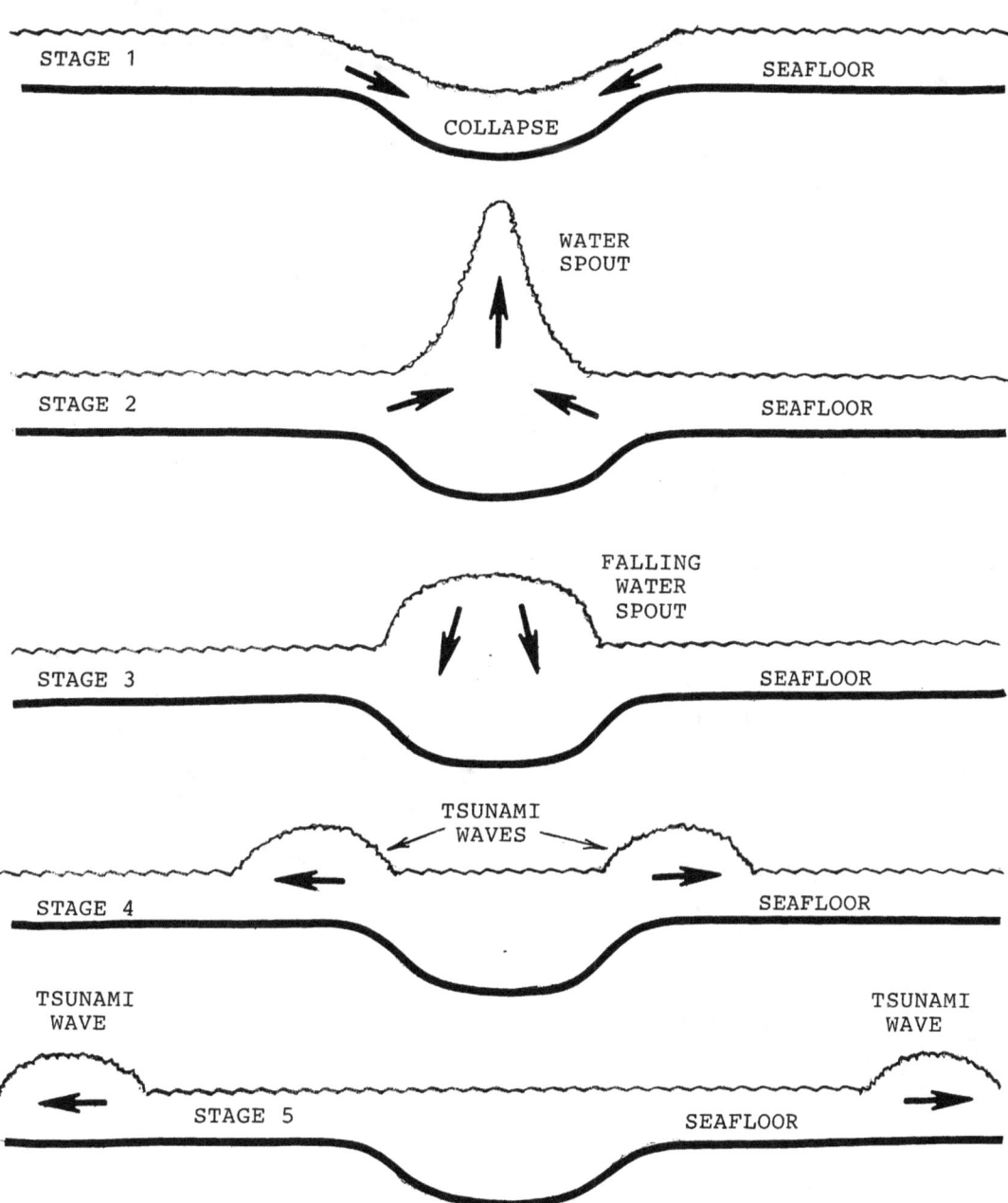

Vertically Exaggerated Cross Sections

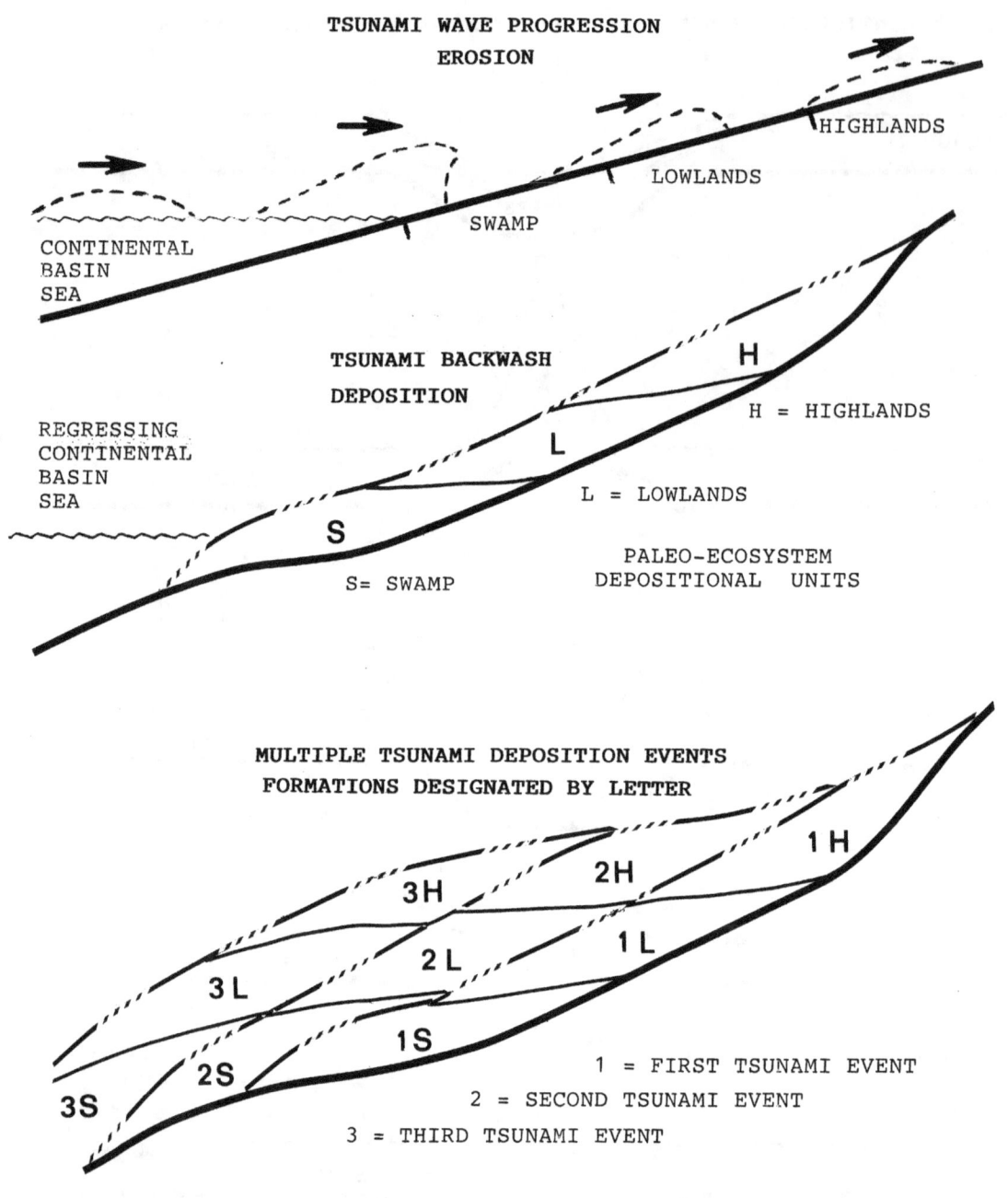

Vertically Exaggerated Cross Sections

    River channels to the oceans eroded deeper, progressively draining continental basin seas. Clastic terrestrial sediment contributions steadily increased, outpacing continental basin subsidence. Each successive tsunami wave deposited similar ecosystem sediments that overlapped previous deposits of the adjacent lower-elevation ecosystem. Mesozoic ecosystem deposits progressively blanketed continental basins from the rim toward the sinking basin center. Each ecosystem was systematically eliminated at the basin center. Laterally-migrating formation deposition produced the observed layer cake succession of Mesozoic ecosystem deposits. Catastrophic tsunami backwash deposition preserved abundant terrestrial fossils and distinctive formation boundaries on continental basin flanks. In contrast to continental basin rims, continental basin centers preserved fewer terrestrial fossils and exhibited less distinctive formation boundaries, due to diminishing tsunami-induced sediment transport.

Cyclothem Deposition

Cyclothems contain multiple sets of thinly-bedded, horizontal layers of differing rock characteristics, which are repeatedly positioned in the same vertical order. Cyclothem deposits are found in Madagascar, New Zealand, Japan, and on every continent (Woodmorappe, 1978, p. 196). However, cyclothems are not forming on any area of Earth today. Traditional models propose that cyclothems formed in delta areas on unstable interior basin shelves that were repeatedly submerged and exposed. However, modern, small-scale deposits along ocean coasts are typically eroded before being consolidated. In contrast, Post-Flood Millennium collapse events generated massive, sediment-laden tsunami sheetflows containing highly-mineralized waters.

Suspended sediments were concentrated in the lower portion of the retreating sheetwash. Sediments segregated by density, shape, and size within a large wave and begin to lag and separate, forming separate slower-advancing waves behind the large leading wave. When the velocity of each trailing wave slows and loses momentum, sediments drop out of suspension sequentually in accordance with density, shape, and size. Each thin, new layer deposited by a slow trailing wave becomes progressively finer-grained as its sediment load is released. Each trailing wave deposits suspended materials overlapping other materials that were deposited by the preceding trailing wave.

Cyclothem sequence layers are frequently skipped. A leading wave may dislodge and carry the uppermost sediments deposited by the previous set of trailing waves. Also, a larger trailing wave segregates and spreads its sediments across a larger area than a smaller trailing wave. Therefore, cyclothem layers are seldom all deposited at any vertical position.

However, the relative order of layers with respect to each other is consistently stacked in the same sequence within each cyclothem set. **Continental basin tsunamis provide the best explanation for Paleozoic and Mesozoic cyclothem deposits.**

Fossils in Tsunami Backwash Sediments

Massive tsunami wave incursions across land provided optimum hydrologic conditions for fossil preservation in clastic sediments. The advancing tsunami wave dislodged terrestrial lifeforms and sediments. The advancing tsunami wave loses momentum, turns, and retreats back toward the continental basin sea. The leading edge of the retreating tsunami backwash churns and mixes plants, animals, and clastic materials in a tumbling motion. That homogenized mixture drops out of suspension, forming a thick sediment layer. Plants and animals buried rapidly within that sediment layer are protected from rapid oxidation decomposition at the surface.

Tsunami backwash burial of plants and animals starts the race between bacterial decomposition and mineral replacement. Slightly alkaline conditions allow most bacteria to thrive and rapidly decompose organic matter. In this case, however, slightly acidic conditions inhibited bacterial growth. Hydrogen sulfide from extruded volatiles in tsunami water produced temporary acidic groundwater conditions. Slightly-acidic, mineral-rich water soaked the ground and recharged through lowland soils and ponds. That mineralized groundwater flowed slowly back toward the continental basin sea through the recently-deposited tsunami sediments. Carbon in bone and plant material attracted and was replaced by silica, calcium, and trace minerals such as uranium. That groundwater flow mineral replacement process rapidly fossilized buried plants and animals. The fossilization process advanced as pore spaces in originally hard animal parts were were infilled by minerals released from groundwater (permineralization). The thickest sediments were deposited at the edge of continental basin seas where tsunami backwash flow subsided. Lifeforms buried within those tsunami deposits were bathed in chemical-laden seawater as tides surged back and forth, facilitating fossilization.

The completion of major porous crust collapse events below continental interior seas signaled the end of the prolific tsunami-induced terrestrial fossilization process.

Coincidentally, swamp and lowland Mesozoic ecosystems captured in fossil form were being vanquished at continental basin centers as those continental basin seas were being infilled and displaced by clastic sediments. **Therefore, Mesozoic ecosystems were unique to Post-Flood Millennium environmental conditions associated with porous crust collapse under continental basin seas and progressive terrestrial sediment infilling from continental basin flanks toward continental basin centers.**

Bones in Fissures

Bones in fissures are evidence of catastrophic Post-Flood Millennium tsunami events. Land animals overrun by massive tsunami waves were violently bounced along the bottom, drowning, crushing, and and tearing apart their bodies. The leading edge of the tsunami wave was cluttered with rolling dead carcasses that were dumped in open rock fissures. Fissure skeletons are too thoroughly broken and randomly scattered for the animals to have died in place. The absence of marks from animal gnawing or human butchering implies a natural calamity. Bones broken but not scarred implies those bones were surrounded by flesh when they were broken.

Most bone breccias in fissures are located on high hills and cliffs. Higher density sediments filled topographic depressions. **Lower density animals, bouncing along the base of the tsunami wave advance, were swept over the tops of hills and dropped into open fissures, one on top of another, until the fissure was filled.** Vegetation floated at the surface of the tsunami wave, avoiding fissure entrapment.

The animal types buried together in fissures are often found today in drastically different climactic zones. However, this common deposit of the past is not forming in modern times. Post-Flood Millennium tsunami transport extended across much longer distances than smaller-scale, modern day tsunami events. The lack of weathering and deterioration of bone fragments implies these un-fossilized bone breccias were deposited in the not-too-distant past! A unique local event is ruled out by worldwide locations of bone breccia.

Near Plymouth, England, fissures in limestone are filled with angular rock fragments and bones of mammoth, hippopotamus, rhinoceros, horse, polar bear, deer, hyena, wolf, and lion. No skeleton is complete. If the animals had fallen into the fissure, the bones would not be so randomly scattered and broken into innumerable fragments. The bones show no sign of wear or gnawing by animals. Similar limestone fissures in Devonshire and Pembrokeshire, Wales, are also filled with broken and splintered bones (Prestwich, 1895, p. 336).

Bone breccias in rock fissures are plentiful in valleys around Paris and on the tops of hills in central France. A 1,430 foot high hill named Mont Genay, near Semur, in Burgundy, is capped by bone breccia remains of mammoth, horse, reindeer, and others. The summit of a flat-topped hill, Mont de Sautenay, near Chalon-sur-Saone, between Dijon and Lyons, has a fissure filled with bone breccia of wolf, cave lion, bear, rhinoceros, deer, horse, and oxen. Numerous clefts in the rocks of the Mediterranean coast of France overflow with bone breccia of diverse animals. The Rock of Gibraltar contains many crevices filled with broken and splintered bones of panther, lynx, caffir-cat, hyena, wolf, bear, rhinoceros, horse, wild boar, red deer, fallow deer, ibex, ox, hare, and rabbit (Prestwich, 1895, pp. 25-48).

Corsica, Sardinia, and Sicily have similar bone beds choking rock fissures. Twenty tons of bones were mined from around one cave of San Ciro near Palermo, Sicily. As with previous examples, the shattered bones are of young and old individuals and show no signs of animal gnawing. Also, the bones are very fresh in appearance, with no indication of weathering or exposure (Prestwich, 1895, pp. 37-38).

A railroad excavation in 1912, near Cumberland, Maryland, cut a closed fissure containing the bones of wolverine, lemming, long-tailed shrew, mink, red squirrel, muskrat, porcupine, hare, elk, peccary, crocodilid, tapir, coyote, badger, puma, beaver, and mink, all covered by gravels (Gidley, 1918, pp. 281-287).

Other bone breccia deposits include Agate Spring Quarry in Sioux County, Nebraska; Big Bone Lick, Kentucky; Kesslerloch, near Thayngen, Switzerland; a gravel pit at Neukoln, near Berlin, Germany; Siwalik Hills in a series of lakebeds along the southern foothills of the Himalayas in India; and the Irrawaddy River deposits of central Burma (Velikovsky, 1955, pp. 67-69, 78-81).

In the village of Choukoutien, near Peking, China, a great mass of bones were found in fissures and caverns. Bone conglomerates included three humans, mammoths, buffaloes, and ostriches (Lee, 1939, p. 370). Most fascinating was Weidenreich's discovery that the three humans were *"a European, a Melanesian, and an Eskimo type lying dead in one close-knit group in a cave on a Chinese hillside"* (Moore, 1953, pp. 274-275). Striated erratic boulders are found in the nearby valleys and on the hills, although this area is considered to have never been glaciated (Velikovsky, 1955, p. 62).

Erratic Boulders

Erratic boulders have been torn from their place of origin, transported across long distances, and dumped on an area underlain by a different type of bedrock. In some specific circumstances, continental glaciers provide the mechanism of transport for terminal (glacier end) and lateral (glacier side) moraine deposits. However, many areas of the world have erratic boulders with no supporting evidence of glacier transport.

Some individuals propose that erratic boulders floated across water on icebergs that melted and dropped the boulders on distant lands. The iceberg must be substantially larger than the transported boulder in order to avoid flipping over or sinking by increased specific gravity of the combined boulder and iceberg. The specific gravity of each boulder exerts a downward force opposite to the iceberg's natural buoyancy.

Ocean water contact soon brings the iceberg's temperature close to thawing. The grip of the iceberg on the boulder rapidly loosens because of downward pressure exerted by the denser boulder against the adjacent ice. Pressure melts ice contacting the boulder by the same physical laws that cause ice to melt under the pressure of an ice skater's blade. The boulder migrates down through the iceberg and drops out the bottom. For large boulders, the potential distance of iceberg transport is extremely short.

The world's largest erratic boulders are found in line on the western shore of the island of Tongatapu, Tonga. These seven house-sized coral boulders are 100 to 400 meters from the coast and weigh up to 1.6 million kilograms. *"The boulders are made of the same reef material found just offshore, which is quite distinct from the island's volcanic soil. In fact satellite photos show a clear break in the reef opposite one of the biggest boulders. And some of the boulders' coral animals are oriented upside down or sidewise instead of toward the sun, as they are on the reef ... "A chain of sunken volcanoes lies just 30 kilometers (20 miles) west of Tongatapu. An explosion collapse of the side of a volcano such as that seen at the famous Krakatau eruption in 1883 could trigger a tremendous tsunami ... Hornbach's analyses of adjacent seafloor topography point to a volcanic flank collapse as the most probable source of such a wave"* (Science Daily, 9/25/08).

Erratic boulders are frequently found in mountain ranges. Either sea level was much higher or the mountains were much lower during transport. Elevation gained between the source area and the final resting place, discredits iceberg transport or slow and gradual tectonic movement.

Loose erratic boulders lying on the Jura Mountains were torn from the Alps as indicated by their mineral composition. The Pierre & Martin boulder is over 10,000 cubic feet. These erratic boulders crossed the Lake Geneva area to reach the Jura Mountains of Switzerland and France. That places those erratic boulders at approximately 2,000 feet above the existing Lake Geneva Valley floor (Velikovsky, 1955, p. 21). This implies the Lake Geneva Valley sank and/or the Jura Mountains rose after these erratic boulders were transported.

Other erratic boulders from Norway were transported across the North Sea area to the shores and highlands of Great Britain. Erratic boulders from Norway also crossed the Baltic Sea and lie high in the Harz Mountains of central Germany. Erratic boulders from Finland were transported across the Baltic area and Poland to come to rest in the Carpathian Mountain area. *"Another train of erratic boulders from Finland stretches across Russia over the Valdai Hills, across Moscow, and as far as the Don"* (Velikovsky, 1955, p. 22).

Long distance erratic boulder transport was probably facilitated by earthquake undulations moving laterally in wave-like fashion across shallow submerged surfaces. Water movement corresponding to laterally-advancing earthquake undulations may have helped to dislodge and roll large round boulders to distant resting places. During the closing stages of the Post-Flood Millennium, mountain upheaval lifted some erratic boulders to their lofty resting place.

Cenozoic Ecosystems

Cataclysmic Mesozoic tsunami deposition transitioned to smaller-scale subaerial deposition conditions during the Post-Flood Millennium. Because continental interior seas were shrinking, Cenozoic formations overlapped Mesozoic formations from continental basin flanks toward continental basin centers. Mesozoic ecosystems progressively contracted until the continental interior sea was infilled with sediments. Modern drainage patterns progressively replaced large continental interior seas. Well-drained "Cenozoic ecosystems" overlapped "Mesozoic ecosystems". Climates became more arid as oceans cooled and continental seas shrank. Porous crust collapse events continued after continental basin seas were displaced. Catastrophic tsunami deposits were replaced by modern subaerial erosion, transport, and deposition processes. The continuing collapse of continental basins during the Post-Flood Millennium maintained a gentle surface tilt toward continental basin centers. That allowed Cenozoic ecosystems to approach and blanket basin centers.

Mountain-building events approaching the close of the Post-Flood Millennium caused frequent subaerial volcanic eruptions. Increased volcanic ash in the atmosphere reflected sunlight and cooled temperatures worldwide. **Mammals migrated to lower-elevation ecosystems as they cooled and became more habitable.** That explains why large mammal fossil remains consistently occupy a high stratigraphic position in the Rock Record.

Mountain Upheaval

The topographic expression of mountain ranges is narrow and sinuous. Mountain ranges appear to have been squeezed by horizontal compression, causing Earth's crust to "wrinkle up". Furthermore, mountain-building apparently occurred during a unique, brief episode of geologic history, without equal in the Rock Record. Ollier and Pain's book, *"The Origin of Mountains"* (2000), examines mountain ranges one by one across Earth's surface. These authors conclude that nearly all of Earth's mountain-building uplift occurred during the late Pliocene and Pleistocene (the most recent two epochs of geological time), and summarize uplift ages in a lengthy table of mountains from all over the world (see Ollier and Pain, 2000., pp. 304-306).

"Uplift occurred over a relatively short and distinct time. Some earth processes switched on and created mountains after a period with little or no significant uplift ... The mountain building period is generally relatively short. It does not appear to be on the same time scale as ... plate tectonics which is continuous. The same rapid uplift occurs in areas where hypotheses such as mantle plumes do not seem appropriate. We do not yet know what

causes this short, sharp period of uplift, but at least the abandonment of naïve mountain building hypotheses might lead to further realistic explanations" (Ollier and Pain, 2000, pp. 302-303).

Most *"mountains and plateaus tend to have very distinctive edges, suggesting uplift of distinct blocks, and to raise such blocks by isostasy alone seems improbable -- there is no way in which erosion alone could cause uplift. There has to be some force within the Earth to push up the land and induce valley erosion, or there has to be a past geological history that leaves a plain out of equilibrium to cause uplift"* (Ollier and Pain, 2000, p. 286).

Low-density volatile components from Earth's interior escaped through the ruptured continental crust following the Worldwide Flood, causing Earth's interior to shrink faster than the surrounding crust. As the increasingly oversized crust adjusted to fit around Earth's shrinking interior, lateral compression was generated, increasing topographic relief. Therefore, Earth's post-Flood continental crust was laterally compressed and predisposed overlap and subduct or buckle upward.

Continental basin porous crust collapse events periodically overpressured crust-forming pyromagma below the base of the solid crust. Low-density pyromagma would not be displaced downward into viscous hypomagma below. Instead, primary pyromagma flow displacement direction was outward at a rising angle, away from continental basin centers, forming a "slip-surface" along that thin, bowl-shaped planar migration zone. Continental basin crust collapse compression squeezed and flattened the uncollapsed porous crust below, causing outward lateral displacement above the bowl-shaped migration zone. "Mushroom Tectonics" occurred as flattened continental basin crust underthrust adjacent crust, generating compressional uplift by forming an overriding wedge.

Mountain range upheaval occurred between separate continental basins or between a continental basin and a stabilized portion of Earth's crust, such as a continental rim, platform, or shield area. The sinuous shape of mountain ranges was generated by the direction of subsurface collapse displacement, timing, and angle of collision between opposing continental basins or butting into stabilized positions of Earth's crust.

MOUNTAIN RANGE UPHEAVAL

MAP VIEW

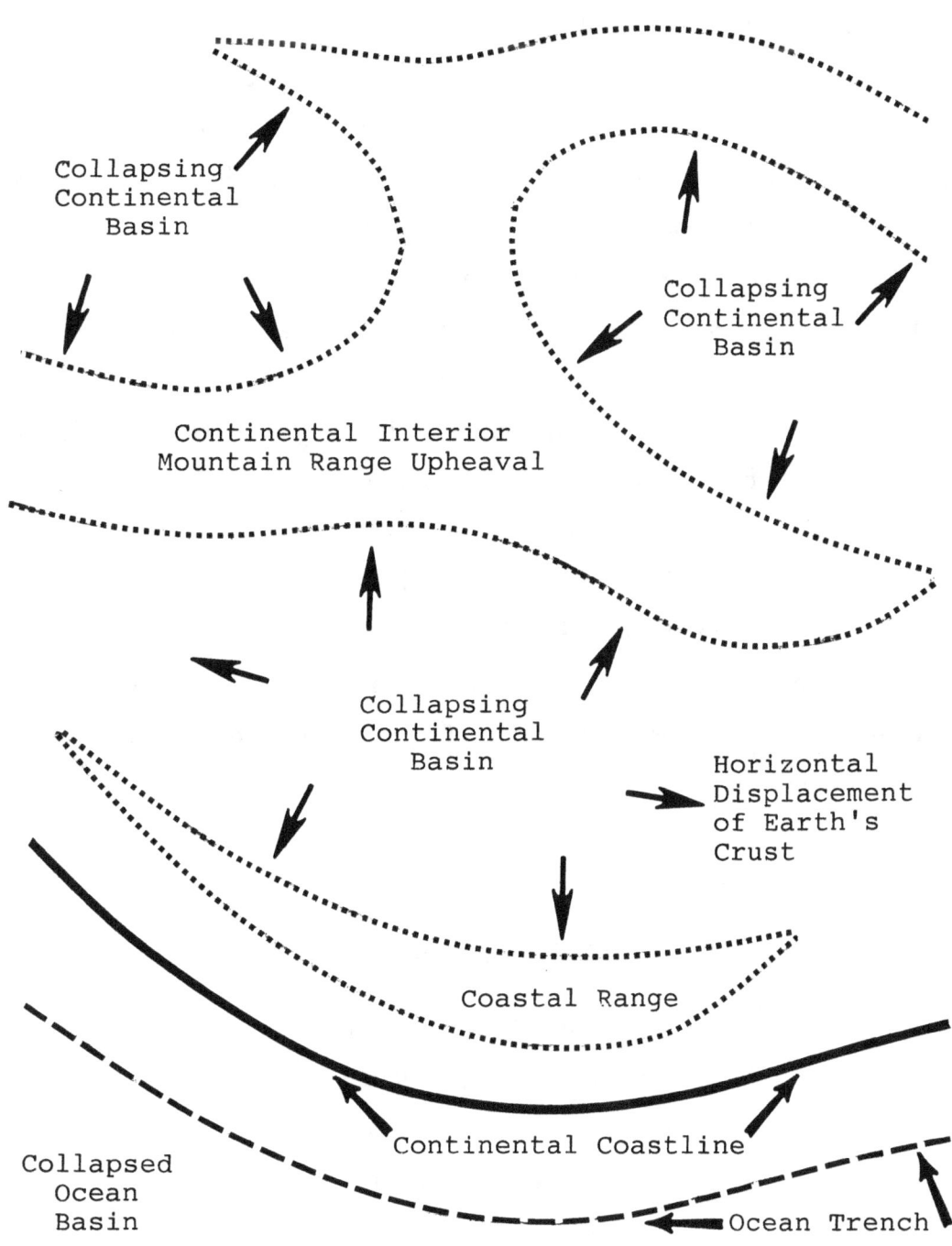

Pyromagma squeezed out from under collapsing continental basins segregated by density and viscosity. Low viscosity components formed channels and traveled faster and further when collapse compression overpressured the pyromagma below. The lowest-density pyromagma components slowly migrated upslope furthest between intermittent porous crust collapse events. Flow dynamics concentrated the lowest density pyromagma components beneath sinuous mountain chains, facilitating thrust slippage of uplifting blocks. Sinuous mountain chains occupy much smaller areas than continental basins. Low-density pyromagma components accompanied by lateral displacement of the base of continental basin curst was focused into much smaller portions of Earth's surface where lateral crust displacement collided to wrinkle up mountain chains.

Mountain range upheaval produced steepening slopes and accelerated erosion. Eroded materials were transported downstream into adjacent continental basins, where they were deposited. Continental basin seas were displaced by the accelerating rate of eroded clastic sediments transported from adjacent rapidly uplifting areas.

During tsunami wave/surface undulation transport, erratic boulders commonly deposited layer of well-rounded boulders, cobbles, and gravel (some from distant sources) at the foot of uplifting mountain ranges. **The final major continental basin collapse events generated the greatest lateral compression of continental crust, causing the greatest isostatic density imbalance, generating the brief late Pliocene and Pleistocene upheaval of Earth's continental mountain ranges.**

Paleogeosynclines

"Ironically, when climbers pose
at the summit of the world's tallest mountain,
they are standing on marine fossils."
(Firstbrook, 1999, p. 51)

The Himalayas are the highest mountain range in the world. Mount Everest towers 29,000 feet above sea level where the air is too thin for birds to fly. Explorers were stunned to discover marine crinoids (Gansser, 1964, p. 164), fossilized fish, and mollusks (Velikovsky, 1955, p. 74) in the rocks of these lofty peaks.

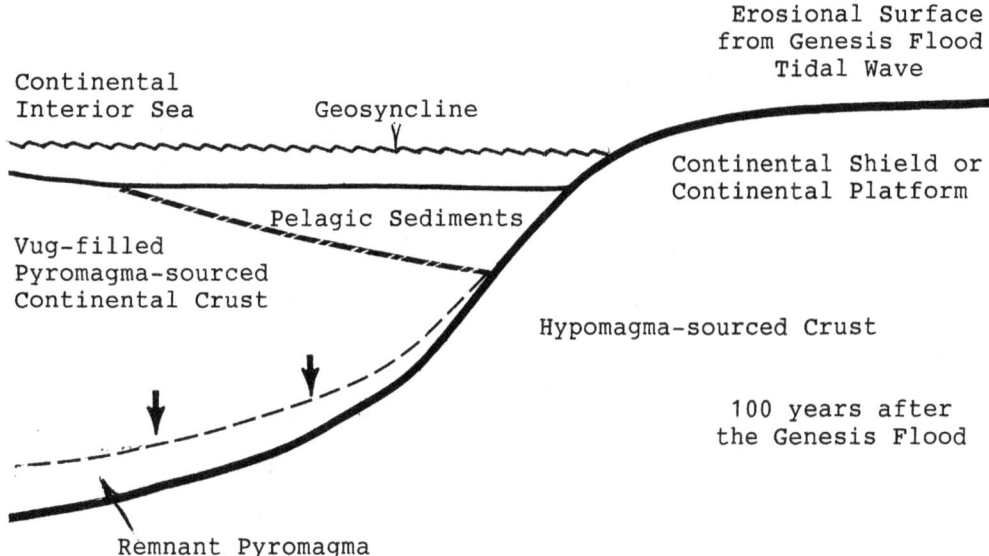

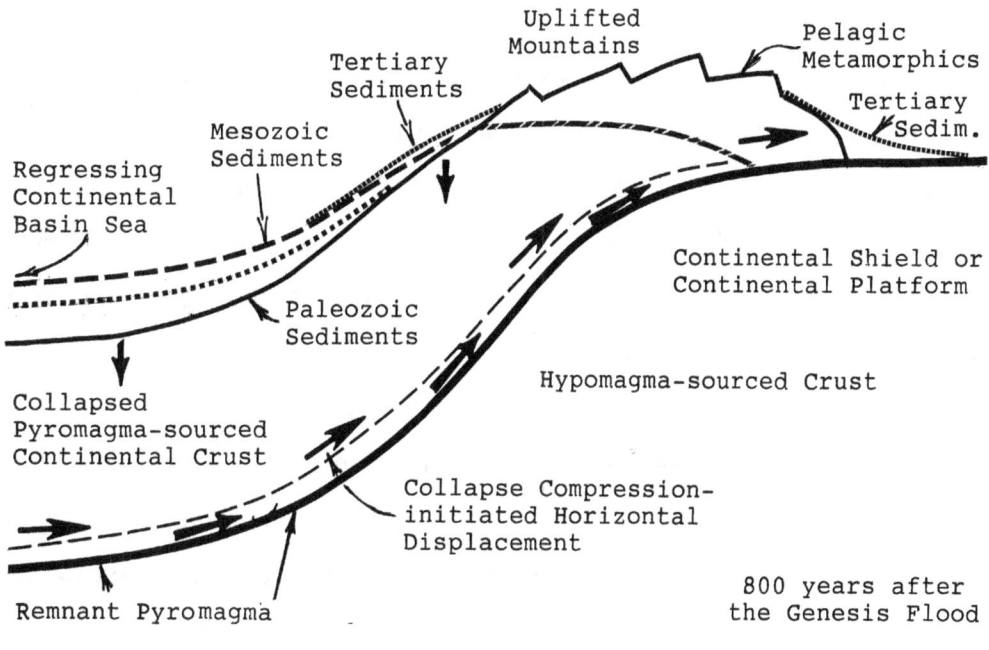

What tectonic mechanism generated a geosynclinal downwarp in Earth's crust and then uplifted those marine sediments to form mountain ranges? During the past century, the definition of a "geosyncline" has been expanded to include all types of sedimentary basins – past or present, oceanic or continental, encompassing a wide variety of shapes, tectonic mechanisms, and depositional characteristics. However, we will focus exclusively on "traditional paleogeosynclines", which were subsequently uplifted to form many of Earth's mountain ranges.

Traditional paleogeosynclines reveal a common set of characteristics which are not duplicated by any modern geomorphic feature. These paleogeosynclines were "trough-shaped" depressions, usually located adjacent to stable shield or platform areas of Earth's continental crust. The absence of land-derived, quartzose sediment, plus abundant marine fossils, imply a depositional environment either far from land or adjacent to a flat-lying, slowly-eroding surface.

The typical absence of cross-bedding, channeling, or current action indicates sediment deposition in calm water below wave base. The absence of algae indicates deposition below the shallow photic zone. The marine fauna is typically characterized by an abundance of pelagic fossils and a scarcity of benthos or abyssal fossils (Selley, 1970, pp. 205-208; Krumbein and Sloss, 1963, pp. 393-396). Therefore, paleogeosyncline sediments were probably deposited in pelagic waters at depths between ten and twenty meters in continental interior seas.

Traditional paleogeosynclines experienced similar tectonic progressions. Tectonic activity fractured rocks along the line of greatest weakness. (Displacement of one side of a fracture relative to the opposite side constitutes a fault.) Fault lines commonly bend around the outside of more stable bedrock. Therefore, tectonic displacement and fracturing following the Worldwide Flood was highly-concentrated around the outside edge of stable continental shield and platform areas.

Extensive linear fracturing along that line of weakness caused the *pyromagma*-sourced porous crust to collapse soon after the Worldwide Flood. Geosyncline troughs were aligned along continental shield or platform borders. **Geosyncline troughs quickly subsided due to extensive fracturing that accelerated porous crust collapse and formed the deepest inland seas during the early Post-Flood Millennium.** Adjacent subaerially-exposed areas shed sediments, progressively filling the geosyncline trough as the seafloor soon collapsed continental porous crust to its base.

As the Post-Flood Millennium continued, intermittent porous crust collapse extended outward from the geosyncline trough on the side opposite from the stable continental shield or platform at a much slower pace across a much wider continental basin area. That porous crust collapse progressively flattened and displaced the continental basin crust laterally towards the adjacent easily-compressible, sediment-filled geosyncline trough. Lateral compression pinched geosyncline sediments against the flank of the adjacent continental shield or platform. The geosyncline trough buckled upward forming a mountain range. That explains why pelagic sediments on continents are commonly restricted to the axis of mountain belts. Subsequent post-Flood collapse caused inland seas to migrate away from rising geosyncline troughs toward adjacent subsiding continental basins.

Paleogeosyncline sediments commonly provide evidence of both volcanic activity and probable metamorphic cementing and corrosion. A *"common characteristic of ancient early geosynclinal pelagic sediments is their association with volcanic activity. This typically consists of diverse pillow lavas, spilites, basalts, and serpentinites"* (Selley, 1970, p. 208). Red nodular limestones associated with several paleogeosyncline facies exhibit scoured bedding surfaces termed "hard grounds", which are rich in iron and manganese and occasionally phosphatized. *"Fossils often show signs of corrosion ... The hard grounds may be due to gentle scouring of early cemented sea beds"* (Selley, 1970, pp. 205-206).

In Kashmir, Helmut de Terra discovered sea bottom sediments at altitudes above 5,000 feet. Bedding planes are tilted as much as 40 degrees from their nearly horizontal deposition orientation, by uplift of the adjacent Himalaya Mountains. To everyone's astonishment, *"these deposits contain paleolithic (recent) fossils"* (Heim and Gausser, 1939, p. 218). This implies the Himalayas were uplifted thousands of feet during the age of man (Velikovski, 1955, p.75). Helmut de Terra concluded the Himalaya uplift continued during the last phases of the Ice Age (de Terra, 1939, p. 222). Swiss geologist, Arnold Heim, concluded that mountain ranges of western China on the east flank of the Himalayas were elevated since the glacial age (Lee, 1939, p. 207).

The ruins of the ancient city of Tiahuanacu are perched high in the Andes Mountains of South America at an elevation of 12,500 feet. *"At the present time, the plateau of the Andes is inhospitable and almost sterile. With the present climate, it would not have been suitable in any period as the asylum for great human masses ..."* (Posnansky, 1945, p. 15). The city of Tiahuanacu was built by highly skilled masons, using enormous stones (Markham, 1910, p. 21). Endless agricultural terraces of the pre-Inca people rise to a modern-day height of 18,400 feet (Velikovsky, 1955, p. 82). Yet today's harsh, high-elevation climate is capable of producing precious little food to support its scant population.

Strand lines of the ancient shoreline of a much larger Lake Titicaca extend south to the edge of the Tiahuanacu ruins. This strand line is tilted, thereby implying the entire area was tilted after Tiahuanacu was constructed. The freshness of strand lines containing modern fossils implies tilting occurred in modern times (Moon, 1939, p. 32).

Chemical analysis was conducted on water from Lake Titicaca and Lake Poopo, and on the salt beds of nearby Coipaga and Uyuni. Several of these lakes and salt beds have chemical composition similar to the ocean today (Posnansky, 1945, p. 23). Archaeological and radiocarbon analyses imply the age of the Andean Culture and of Tiahuanacu cannot be much older than 4,000 years (Hibben, 1951, p. 56). The evidence implies Lake Titicaca and the Andes Mountains were uplifted from a near sea level position to its almost uninhabitable elevation of over 12,000 feet after the Inca civilization flourished.

The raised beaches of Valparaiso, Chile, reveal a surf line at an elevation of 1,300 feet above sea level at the foot of the Andes. Sea shells at this altitude were whole and perfectly preserved, implying recent uplift. Because only a few intermediary surf lines can be detected, uplift of the area could not have proceeded little by little (Darwin, 1835, Chapter 15).

Mushroom Tectonics

Continental basin porous crust collapse was an intermittent top-down progression. The final mountain building collapse events generated outward horizontal displacement deep in Earth's crust. That produced a form of thrust faulting known as "mushroom tectonics". Examples include the southern Rocky Mountains, the Colorado Plateau, the Uinta Plateau, and the Andes of Ecuador (Ollier and Pain, 2000, p. 160).

MOUNTAIN RANGE UPHEAVAL PROGRESSION

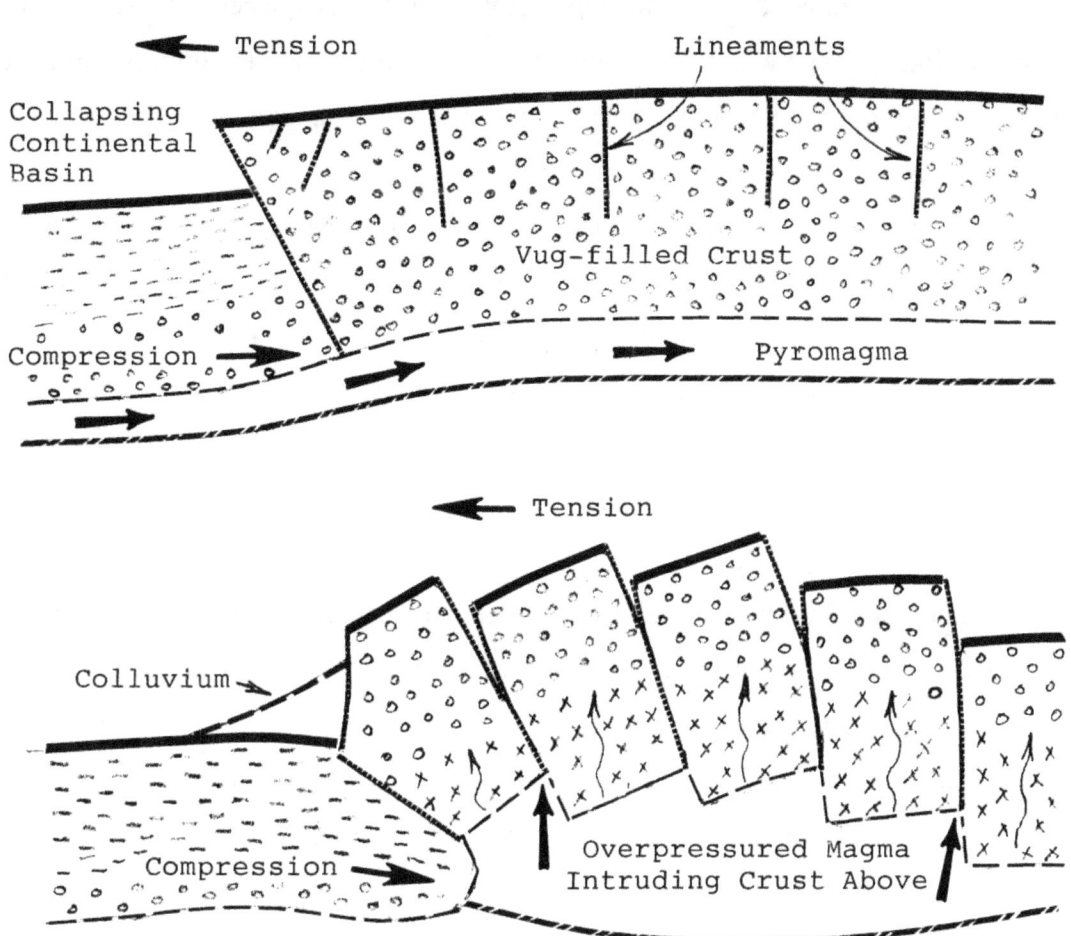

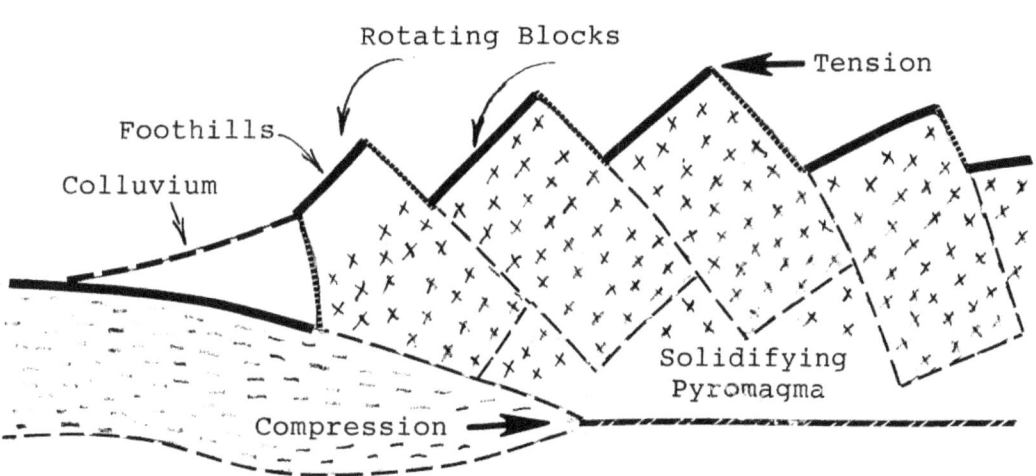

Vertically Exaggerated Cross Sections

An uplifting mountain range is squeezed horizontally from one or more directions. Thrust faults dip in the direction of advancing lateral subsurface displacement. The uplifting block is shaped like a wedge with the wide end up. **Horizontal compression undercuts and thrusts up the wedge-shaped block, forming mountains. Consequently, the "roots" of mountains narrow with increasing depth.**

The uplift of a wedge-shaped block produces an "overhang" above the adjacent plains. As mountains raised, some of them spread laterally, thrusting rocks over the surrounding lowlands (Ollier and Pain, 2000, p. 12). The overhanging portion of the uplifting mountain block lacks support, splits (fractures by tension), and collapses onto the top of the adjacent plain. This wedge-edge collapse tilts the collapsing portion of the uplifting block toward the adjacent plains. Sedimentary layers and foliation planes that were oriented nearly horizontal prior to uplift, now dip steeply toward the adjacent plains.

Precambrian metamorphic foliations are herein advocated to primarily be mineralized bedding planes formed by vertical collapse of porous crust. The vertical compression of vugs generated horizontal fractures as the vugs were crushed. Encapsulated volatiles were forced out of their collapsing vugs into a planar horizontal surface, where they solidified at various mineral melting points, depositing their mineral signatures in the form of bedrock foliations. I have taken thousands of strike and dip readings at hundreds of locations on metamorphic rock foliation planes in the mountains west and southwest of Denver. The most common foliation plane orientation dips to the northeast. The adjacent plains are to the east-northeast in this area. Assuming original horizontal orientation, the observed foliation plane dip orientation supports the concept of wedge uplift followed by "overhang" collapse to the northeast.

Metamorphism of Mountains

During the final stage of the Post-Flood Millennium, sinuous mountain chains were uplifted by lateral compression displacement. Remnant *pyromagma* was squeezed by the collapsing crust above in "rolling pin" fashion, outward away from continental basin centers, toward positions under adjacent uplifting mountains. ***Pyroclastic* volatiles were over-pressured by horizontal compression of the crust and injected upward through**

fractures. Overlying sedimentary and porous crust bedrock was metamorphosed by those rising fluids and gases. Associated continental volcanic activity increased to its zenith eight to ten centuries after the Worldwide Flood.

The low-density hydrous silicon dioxide fraction of *hypomagma* was rapidly injected into open lineament fractures and rose toward the surface. However, that hydrous *hypomagma* became extremely viscous, upon contact with cooler bedrock, and solidified as pegmatite in open fractures. Large *hypomagma* chambers, known as "batholiths", rose more slowly, melting and absorbing the crust above, as they cooled and solidified.

Outcrops of mountain cores frequently contain minerals, like coesite, that solidify under conditions of high-pressure and low-temperature (Smith, D.C., 1984, pp. 641-644; Chopin, 1987, pp. 183-197; Wang, et al., 1989, pp. 1085-1088; Hsu, 1991, pp. 107-110; Shutong et al., 1992, pp. 80-82). The rapid, periodic compression and upheaval associated with continental basin collapse induced mountain building events and explains the high-pressure, low-temperature conditions needed to form coesite.

Continental Basalt Flows

Massive basalt flows cover sizeable portions of continents. The most curious feature of continental basalt flows is the restricted times of extrusion. Sedimentary deposits interbedded with basalt flows are dated from "Precambrian" to "Cambrian" and from "Mesozoic" to "Tertiary". With good reason, continental basalt flows were not extruded during the "Paleozoic" and "Quarternary" sequence (according to the Geologic Time Scale).

So-called Precambrian/Cambrian *basaltic magma* flows are exhibited in the North Australian Province and in the Keweenan Province on the Canadian-American border near Lake Superior. The initial Precambrian/Cambrian phase occurred as Earth's crust was horizontally compressed during the Regressive Stage of the Worldwide Flood as collapsing ocean basin crust displaced laterally and underthrust continental crust. The continental crust uplifted, flexed, and ruptured, causing over-pressured *basaltic magma* to extrude onto the land. In contrast, "Paleozoic" sediments were sourced by the escape of *pyromagma* volatiles from vertically-collapsing porous crust below. Because the porous crust underlying continental basins was predominantly of *pyroclastic* composition, basaltic magmas were rarely if ever available for volcanic extrusion in those areas.

The second "stage" of *basaltic magma* flows occurred during mountain upheaval at the end of the Post-Flood Millennium. These so-called "Mesozoic" and "Tertiary" *basaltic magma* flows occurred on the Siberian Platform of the Soviet Union; the Karoo Province of South Africa, Zambia, Namibia, and Mozambique; the Parana Basin of Brazil, Uruguay, Paraguay, and Argentina; the Deccan Plateau of India and Pakistan; the North Atlantic Province of Greenland, Baffin Bay, Northern Ireland, and Norway; the Ethiopian Province of Ethiopia and Yemen; the Eastern China Province; and the Columbia River Province of Washington, Oregon, and Idaho.

The Columbia River Province exhibits at least 120 major flows (Tolan, et al., 1989, pp. 1-20), extruded during lateral crust compression associated with mountain upheaval events. The Columbia Plateau lava blankets two hundred thousand square miles in Washington, Oregon, and Idaho. The Snake River cut Seven Devils Canyon to a depth of over three thousand feet without reaching the bottom of the lava flow (Chamberlin, 1937, p. 85). *"All competent observers have remarked about the freshness* (unweathered appearance) *of lava deposits in the Snake River Valley in Idaho"* (Wright, 1911, p. 688).

Quarternary sedimentation does not contain large scale *basaltic magma* flows because the last catastrophic "Tertiary" mountain-building events had been completed. **Large-scale *basaltic magma* flows were only expelled when lateral compression of Earth's crust was temporarily at its maximum. *Basaltic magma* flows occurred 1.) during the Regressive Stage of the Worldwide Flood and 2.) in association with intense mountain-building at the end of the Post-Flood Millennium.**

Continental Rift Systems

Due to their varied locations and shapes, continental rifts are not simply explained. The North American Midcontinent Rift System (MRS) is proposed to be thousands of kilometers from the edge of any tectonic plate. In contrast, the Great African Rift (GAR) is sometimes considered to be a tectonic plate boundary. MRS exhibits an arcing "hook" shape. In contrast, GAR is a generally linear hinge-like feature.

Similarities between MRS and GAR include a tensional pulling apart of Earth's crust, and multiple intermittent basaltic magma extrusions flattening the rift valley floor. The location of continental rift valleys is typically distant from continental boundaries.

CONTINENTAL RIFT FORMATION

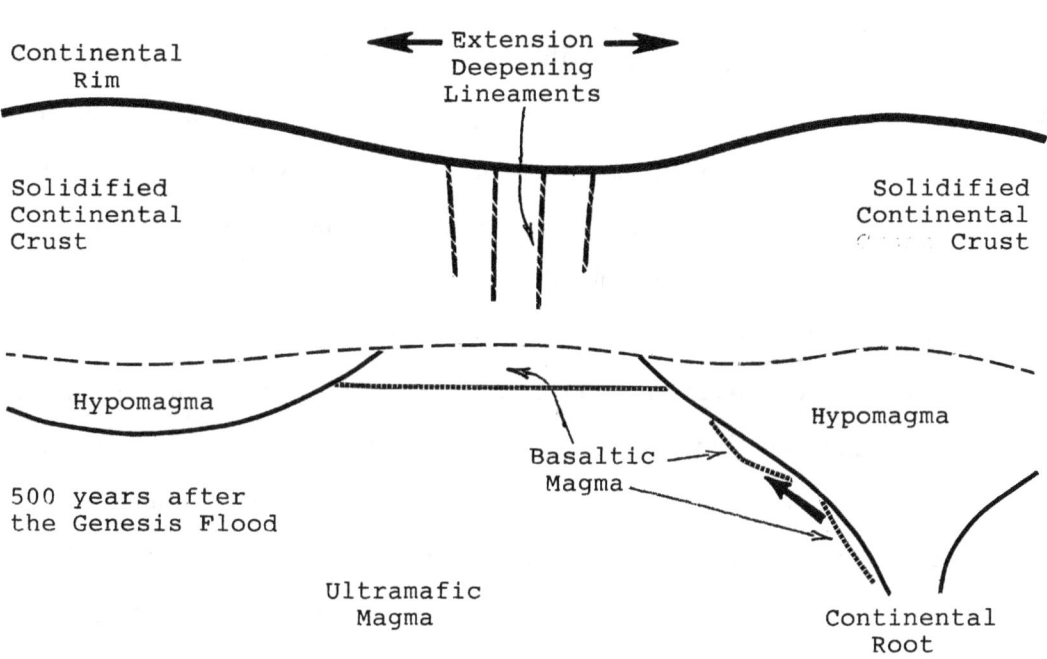

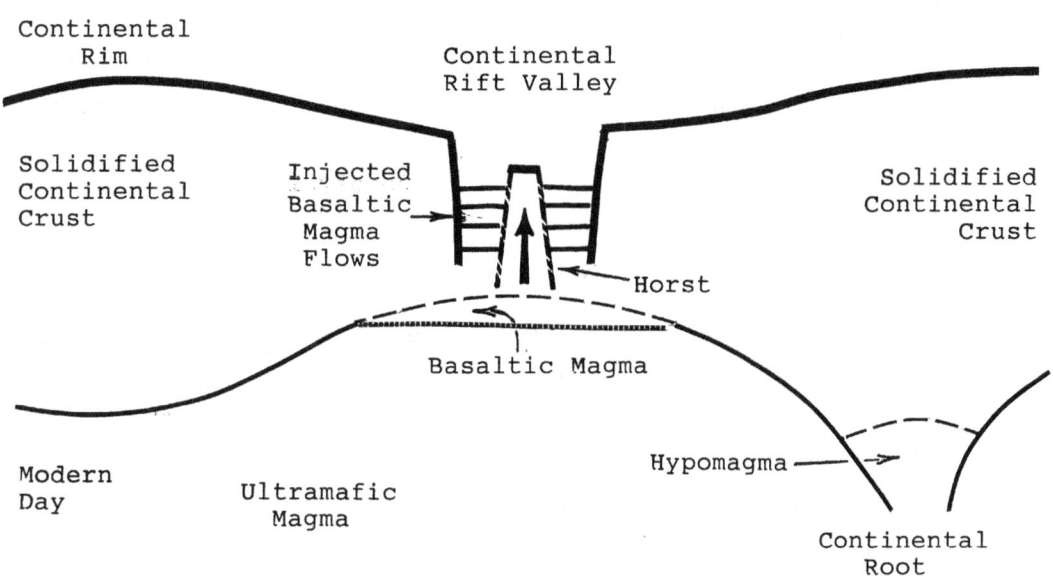

Continental interior degasification caused gentle undulations in Earth's continental crust during the Post-Flood Millennium. The lower density fraction of remnant *basaltic magma* migrated updip to occupy areas under undulation crests. The resulting upward "bowing" stretched the surface above, generating tension along the crest of the undulation. Lineaments along the thinnest portion of stretched crust tore open to form a continental rift valley.

Basaltic magma ascended in the rift valley to its isostatically stable position. However, that isostatically stable position was progressively destabilized as the undulation grew more pronounced and low-density *basaltic magma* slowly migrated under the undulation crest. Consequently, the tear was periodically reopened as the basaltic magma underplate gradually thickened and isostatically lifted the gap in the overlying crust. Each new *basaltic magma* extrusion was sequentially deposited on top of the previous basaltic magma flow. However, the basaltic magma flows never filled the rift valleys to their rims, because the crust flanking the rift was of lower density than the basaltic magma intrusions.

Separation of the African continent occurred along lineament fissures, that tore apart to form the Great Africa Rift, extending across approximately one-sixth of Earth's circumference. Modern marine fossils and sharp fault escarpments imply great Earth movements at a recent date (Gregory, 1896, pp. 5, 236).

The North American Midcontinent Rift System (MRS) is a major Precambrian basement tectonic feature running through portions of Kansas, Nebraska, Iowa, Minnesota, Wisconsin, and Michigan, U.S.A. *"In cross section the MRS commonly consists of a thick medial horst of extrusive and intrusive (igneous) rocks surrounded and covered by basins filled with epiclastic and apparently non-marine clastic sedimentary rocks"* (Reed, 2000, p. 162).

"The MRS is a large intracratonic rift – an extensional crustal feature that locally penetrated the entire thickness of the continental crust. During the event that produced the MRS, structural downwarping and normal faulting were accompanied by the extrusion of tremendous volumes of magma as flood basalt flows ... Late in the MRS event, tectonic inversion resulted in reverse motion along the original normal faults. The resulting structural uplift created a central horst with flanking half-graben basins in the western limb of the MRS" (Reed, 2000, p. 166).

The central horst along the west flank of the MRS behaves like a plug in the neck of a volcano. *Basaltic magma* over-pressuring causes the central horst to gradually rise; weakening fractures between the central horst and adjacent valley floor basalts. Rupture and expulsion of basaltic magma through adjacent fractures, allows the central horst to sink and reseal those fractures.

Clastic Dikes

Vertical "walls" of sandstone that have been injected upward into a surrounding clay layer are known as "clastic dikes". What caused the sand to flow? How did the sand penetrate the clay layer above and sometimes erupt as muddy sand volcanoes? Why was the overlying clay layer torn apart and intruded with sand?

Pressurized water rising from below at a rapid rate destabilized the sand grain packing arrangement. That "quicksand" condition allowed sand to flow freely. Fluid over-pressuring from below is commonly referred to as "earthquake liquefaction".

Porous crust collapsed far below the unconsolidated sand layer. That collapse shattered vugs within the rock matrix, releasing volatiles including water, dissolved carbon dioxide, and mineralized fluids. The area above the collapse zone sank. Overpressured volatiles flowed upward through collapse-induced fractures. Ascending liquid carbon dioxide "flashed" into carbon dioxide gas as pressure decreased approaching the surface. Rising fluids and expanding gases raced into the unconsolidated sand layer above, over-pressuring and destabilizing the sand grain packing arrangement.

A clay layer above briefly acted as a permeability barrier. However, the over-pressured sand layer below swelled, stretching the clay layer above until it ruptured along its weakest vertical plane. Free-flowing sand ascended through the rupture and spouted out onto the surface as a muddy sand volcano. **Hot water, steam, carbon dioxide gas, and mineralized fluids vented through the sand-filled rupture to the surface, metamorphosing the "wall-shaped" clastic dike into sandstone.**

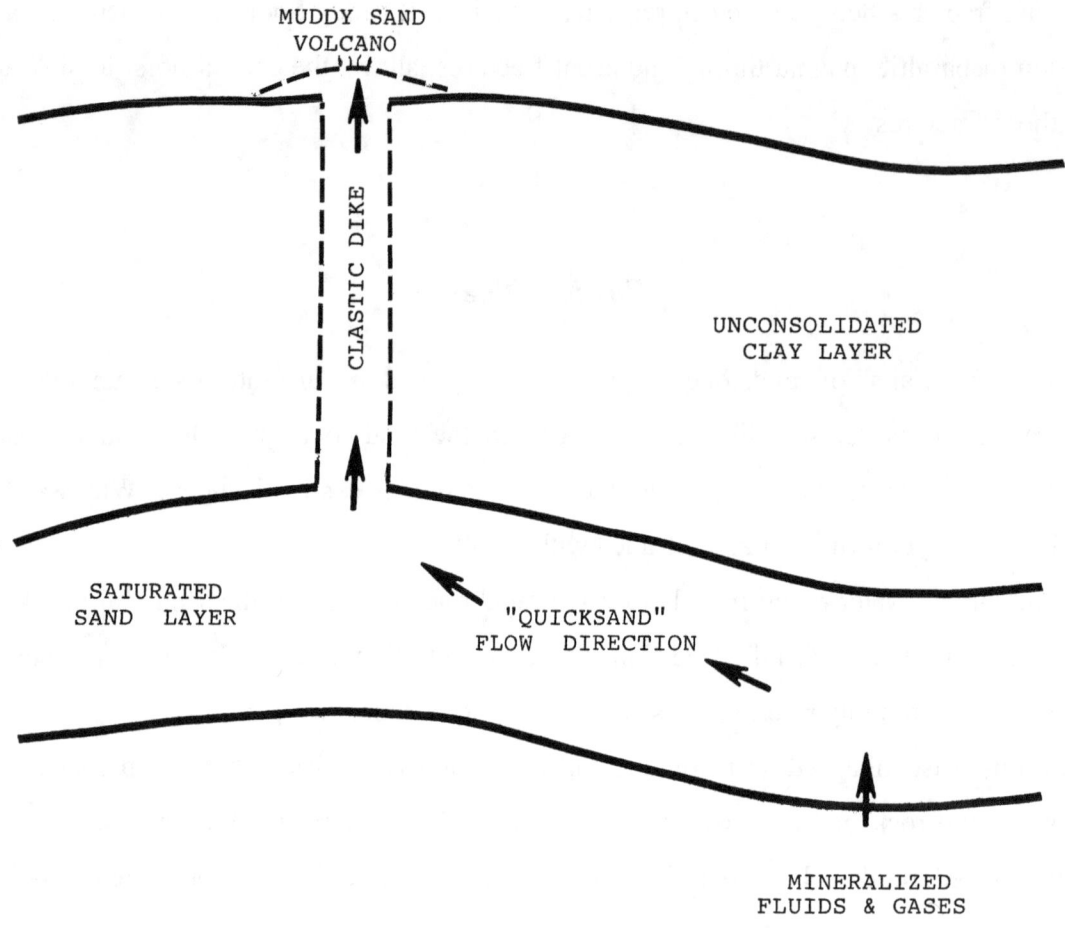

Batholiths

"Batholiths are generally discordant, and most consist of multiple intrusions. In other words, a batholith is a large composite body produced by repeated, voluminous intrusions of magma in the same area" (Monroe and Wicander, 1994, p. 61). *"Recent geological and geophysical observations have revealed that the world's granitic plutons are mostly tabular in shape and typically only a few kilometres thick"* (Walker, 2007, p. 15; also see Clemens,

2005, p. 14). *"The long-cherished picture of granitic diapirs [balloons of magma] slowly pushing their way toward the upper crust and grinding to a halt by solicification has been replaced by an altogether different picture of narrow feeder dykes punching their way upward in months, pulsing with magma and feeding rapidly growing plutons"* (Clemens, 2005, p. 15). *"Furthermore, for magma to ascend to the surface it is found that the critical widths of the dykes are quite small, of the order of 1-2 m only. In other words, narrow dykes can be very efficient transporters of granitic magma in the crust"* (Walker, 2007, p. 14; also see Clemens, 2005, p. 13),

"Clemens describes how the crystals in some granites are arranged in patterns resembling textures in sedimentary rocks: graded-layering, cross layering, scour and fill structures, flame structures and swirls or enclaves of crystals (see Clemens, 2005, pp. 9, 10) ... They point to the fact that the magma was flowing when the crystals settled, and that the flow was pulsing. These support the concept that the batholiths filled quickly during times of tectonic disturbance" (Walker, 2007, pp. 14-15).

Textured patterns indicate the magma was intruded as a crystal melt mush. Minerals in magma crystalize in sequence according to their unique temperatures of solidification. The rapid pulsing transport of granitic magma between the source and batholith would have continually mixed the liquid components of the crystal melt mush, restricting net crystal growth during transit. It is reasonable to conclude that the crystal melt mush consistency of the magma was present at the pre-transit source.

Earth's interior would have long since been stabilized if the source zone is assumed to be billions of years old at the time of granitic magma transport. From that perspective, secondary heating would likely have melted expelled select minerals during granulite metamorphism. Expelled minerals would supposedly congregate at the granitic magma source. However, excess water would have been previously expelled when the rock originally solidified. *"The mineral transformations that occur during metamorphism are the result of chemical reactions, and these need abundant water to allow the free exchange of ions"* (Walker, 2007, p. 14). If the granitic magma was derived from heating and selective mineral expulsion during granulite metamorphism, model parameters must account for an independent water source and an independent heat source.

In contrast, Collapse Tectonics proposes Earth was less than three thousand years old at time of magma migration. The crust was continuing to solidify deeper. Granitic magma was sourced from the crust-forming magma zone. Excess water would have been abundantly present in the crust-forming crystal melt mush source zone at a depth between 20 km and 30 km. Excess water would have provided low-viscosity flow capability, greatly accelerating upward transport to rapidly-fill the batholith. *"Viscosity calculations* (see Clemens, 2005, p. 13) *have shown that the flow properties of granitic magma remain relatively unaltered by the presence of crystals"* (Walker, 2007, p. 14).

"So how long does it take for magma to ascend 20 km in the crust? With typical magma and crust properties it could be anywhere between five hours and three months ... What sort of time would it take to build a huge pluton? According to Clemens, a dyke 3 m wide and 1 km long (in plan) could build a batholith of 1,000 km^3 in 1,200 years ... "Given this tabular shape, it is a simple matter to model the cooling by conduction of a 3 km sheet of granitic magma. Based on conduction alone (i.e. ignoring the cooling effect of fluids) it would take only 30,000 years to completely solidify from the initially liquid magma. But we know that fluids play a controlling role in the cooling of granitic magma, and their behavior would drastically reduce the time" (Walker, 2007, pp. 14, 15; also see Snelling and Woodmorappe, 1998, pp. 42-44.)

Batholith emplacement was prevalent during the final centuries of the Post-Flood Millennium. Porous crust collapse events were generating subsurface displacement of crust outward and upward away from continental basin centers. That caused mountain ranges to wrinkle up in adjacent areas with each porous crust collapse event. Low-density granitic magmas at 20 km to 30 km of depth migrated outward away from subsiding continental basins and upward under uplifting mountain ranges.

Compression-driven uplift and isostatic adjustments routinely fractured the overlying crust. Select water-rich magma components from temporarily-overpressured batholiths intruded upward through recently opened fractures above. Intruded sediments commonly exhibit partial melting. Hydrothermal alteration occurred when expelled magmatic fluids "bathed" the host rock. The most recognizable fracture intrusives are white, pegmatite (hydrous quartz) veins and dikes that cross-cut darker host rocks.

Water *"has a very high heat capacity, a measure of the ability of a substance to absorb heat. One gram of H2O can absorb much more heat than one gram of rock"* (Young & Stearley, 2008, p. 326). Mountain snowmelt and rainfall recharge the fracture system. Groundwater descends, picks up heat from the batholith below, and flows along a steeply tilted water table toward lower elevations where it is resurfaces later in springs, streams, and rivers. Hot mineralized water expelled from the solidifying batholith mixes with surface-sourced groundwater and rises to the surface, being expelled as a hot spring. **Most batholith emplacement occurred during the final centuries of the Post-Flood Millennium in association with mountain range upheaval. Batholith cooling and solidification was completed within a few thousand years thereafter.**

Calderas

Porous crust collapse compression events intermittently squeezed a volatile-rich fraction of remnant *pyromagma* beneath continental basins up-dip along the underside of the porous crust away from continental basin centers. If the slope of the overlying crust flattened or reversed dip in all directions, the pyromagma intrusive layer accumulated, causing the crust above to isostatically rise and fracture. Pressures increased as pyromagma accumulated and volatile liquids expanded into gases. The crustal seal above was progressively weakened and broken, unleashing an explosive *pyroclastic* eruption. Volcanic ash was blown into the upper atmosphere. The large-volume expulsion of volatiles occupying the pyromagma chamber below caused Earth's crust above to sink. That formed a caldera depression, typically ranging from ten to thirty kilometers in diameter.

As soon as the volcanic eruption completed its release of over-pressured volatiles, remnant magma minerals solidified, sealing open fractures. Continuing *pyromagma* migration resupplied the magma chamber, which may or may not shift laterally from its original position. Pressures increase preceeding the next caldera eruption cycle. However, pyromagma sources progressively deplete. The refill rate of the expanding *pyromagma* chamber slows accordingly. When *pyromagma* chamber recharge equals chamber leakage, caldera eruptions cease.

Three overlapping calderas have been identified in Yellowstone National Park in northwest Wyoming, U.S.A. According to the secular Geology Timescale, these caldera-producing eruptions occurred *"2 million, 1.3 million and 640,000 years ago"* (Farquhar, 2007, p. YJ-13). **In contrast, Collapse Tectonics proposes these three caldera events occurred approaching the end of the Post-Flood Millennium and thereafter.** Gradual localized uplift indicates the depleting *pyromagma* source is still active. However, chamber leakage may be sufficient to prevent or greatly minimize a future caldera eruption.

Limestone Caves

Immediately following the Worldwide Flood, large portions of continental interiors were inundated by shallow seas. Post-Flood Millennium collapse of porous crust released abundant calcium carbonate-rich volatiles. Extrusion of calcite-rich volatiles into seawater above produced widespread deposition of calcium carbonate mud. Gases were released, forming "bubbles" within the rapidly-solidifying limestone.

As subsiding continental basins filled with sediments, modern outlet rivers breached the flanks of basins, progressively draining continental seas. Water tables dropped accordingly. Subsequent porous crust collapse events expelled abundant hydrogen sulfide and fractured overlying limestones. Contact with groundwater produced sulfuric acid-rich fluids and hydrogen gas. Overpressured sulfuric acid-rich fluids were injected up collapse-induced fractures. Rising acidic fluids dissolved and absorbed calcite from fractured limestone sidewalls to form vertical caverns. Groundwater flows containing high concentrations of sulfuric acid probably dissolved most limestone caverns (see Polyak and Provencio, 2000, pp. 72-79).

The groundwater table is the contact between free-flowing water below and air above filling bedrock fractures and permeable voids. Rain and snowmelt descend through soils and fractures, "recharging" the groundwater table. Beneath topographic highs, groundwater tables slope laterally downward toward lower surface elevations.

Overpressured hydrosulfuric acid solution rose to the level of the local groundwater table, turned and flowed laterally through the maze of fracture conduits and was expelled into surface streams and rivers. Caverns were dissolved in limestones and dolomites all along that route, which conforms to the observed cavern arrangement of vertical shafts intersecting with gently-sloping tunnels.

Ocean waters, which were warm at the end of the Worldwide Flood, gradually cooled, slowing evaporation. Climates became more arid. Continental basin seas dried up.

Groundwater tables dropped below many caverns, setting the stage for the decorative phase of limestone precipitation in caves.

Rain absorbs carbon dioxide, producing a very weak carbonic acid solution. Calcium carbonate in limestone strata is slowly dissolved by weak carbonic acid. Calcium carbonate-rich groundwater percolates down into limestone caverns above the deeper modern water table. Warm cavern temperatures induce calcium carbonate precipitation, forming decorative stalactites, stalagmites, and flowstone.

Carolina Bays

Carolina Bays are a slow-developing continental variation of orthogonal seafloor topography. Surface expression exhibits closely-spaced marshy elliptical depressions along the Carolina coast. Similar depressions are present across the Atlantic Coastal Plain from New Jersey to Florida. The number of these bays is estimated between tens of thousands and half a million (Johnson, 1942: Prouty, 1952, pp. 167-224). The more prominent bays average 2,200 feet (0.67 km) in length. One remarkable feature of these bays is their striking parallelism, with the long axis oriented NW-SE.

"The optical dating results indicate that present day bay morphology is not the result of a single event, catastrophic formation, but rather they have evolved through multiple phases of activity and inactivity ... This is evidenced by multiple rims of differing ages along the same bay, and by multiple ages within single rims" (Ivester et al., 2004, p. 69). Shallow water tables and the absence of surface drainage are indicative of slow progressive deposition from windblown (eolian) transport plus organic contributions from the local habitat. Gradual sediment accumulation began during or preceding the Ice Age and continues into modern times.

The most fascinating characteristic of Carolina Bay's is their narrow white sand rims. The white sand rims are commonly elliptical and less frequently circular in shape. The highest portion of the white sand rim is typically located at the southeast end of each Carolina Bay (Velikovsky, 1955, p. 98).

Lineament intersections provide the most effective fracture conduits for rising groundwater flow from deep in Earth's crust. However, in areas where lateral torque is present in the crust, one fracture extending away from each fracture intersection will consistently widen in the same direction. Lateral torque appears to have been operating, as evidenced by the elliptical orientation of the long axis of Carolina Bays systematically rotating along the Atlantic coast from N16W in east-central Georgia to N64W in Virginia (Kaczorowski, 1977). In these areas, torque-widened fractures characteristically extend to the northwest from each fracture intersection. Escaping volatile flow at the surface was greatest at the fracture intersection and progressively declined outward along the torque-widened fracture, extending to the northwest.

Hot springs are characteristically supplied by mineralized water contained in "mineral precipitate pipes". By comparison, "precipitate permeability plugs" typically form where a brief mineral-rich flow was followed by an extended period of negligible flow. Groundwater flow in this flat-lying coastal plain was too slow to flush minerals away prior to precipitation. Extruded volatiles cooled and precipitated out, infilling granular porosity to form "precipitate permeability plugs" in overlying sediments.

Initially, escaping mineralized fluids rose vertically from each fracture intersection. Subsequent collapse events intermittently extruded mineralized fluids that flared outward and upward in elliptical "cone-shapes" as they ascended beneath enlarging "precipitate permeability plugs". As overburden sediments slowly accumulated, the surface position of elliptical cones of ascending mineralized groundwater transport progressively widened approaching the surface.

Periodic pulses of upwelling mineralized water generated temporary destabilization of the mineral grain packing arrangement. That produced a "quicksand" condition along the more permeable "cone-shaped" conduit plane flaring outward and upward beneath the sides of each "precipitate permeability plug".

The relative density of "permeability plugs" increased as minerals displaced water in intergranular pore spaces, rendering the mineralized plug to be slightly denser than the surrounding sand. **Volatiles rising from below destabilized the sand packing arrangement, allowing the denser mineralized plug to sink and form a characteristic Carolina Bay surface depression.** Lower-density destabilized sand flowed up and outward, expelling the elliptical narrow white sand rim at the surface around the outside perimeter of the cone-shaped mineralized plug.

The fracture intersection released more volatile flow than the widened fracture extending to the northwest. Therefore, the highest sand rim typically formed at the southeast end of the elliptical feature, due to a higher flow velocity. The comparatively clean white sand rim was produced by rapid intergranular flushing of finer-grained and organic materials during each collapse-induced mineralized fluid extrusion event. The northwest-trending elliptical shape of the white sand rim was formed in accordance with the linear extension of mineralized fluid release at depth.

Mineral precipitate plugging infrequently blocked fracture conduit transport, altering subsurface release positions of rising volatiles into unconsolidated sediments above. In those rare cases, the ensuing hydrologic process shifted laterally to form an overlapping elliptical Carolina Bay feature. The varying carbon age dates of the overlapping rims correspond to the age of the organic materials being flushed through the white sand conduit.

Collapse Tectonics predicts that Carolina Bay depression sediments will exhibit lower permeability and higher density (due to mineralized plugging) than unconsolidated sediments flanking the depressions. The white sand rims should exhibit the highest permeability and dip from the surface downward at an angle toward a position beneath the Carolina Bay depression center.

Ecosystem Migration

Continental basin seas were progressively displaced by sediment infilling sourced by adjacent mountain range upheaval approximately 800 to 1,000 years after the Worldwide Flood. Paleocene, Eocene, Oligocene, Miocene, Pliocene, and Pleistocene Epochs of the Tertiary Period correspond to successively colder and higher-elevation ecosystems that produced new ecological niches, which came into existence one after another as Earth's mountain chains periodically rose, leading to ecosystems populated by specialized lifeforms capable of adapting to increasingly colder, drier, higher-elevation climates.

These Tertiary Epochs do not represent exclusive time slots in geologic history. Instead, these ecosystems developed and migrated downslope from uplifting mountain ranges as the local climate became colder and drier. This downslope migration of Tertiary ecosystems generated an over-lapping stratigraphic sequence of sedimentary formations.

Uplifting mountains steepened slopes, accelerating sediment erosion, transport, and deposition rates, preserving these overlapping Tertiary ecosystems in sequence. Meanwhile, Tertiary highland formations were overlapping Mesozoic lowland formations, which were overlapping Paleozoic marine formations. The lateral ecosystem march toward sinking continental basin centers successively eclipsed marine, then lowland, then highland ecosystem deposition as sediment influx from uplifting mountains overwhelmed basin subsidence. Early post-Flood marine ecosystems were vanquished centuries before Tertiary ecosystems became established. **However, Phanerozoic sedimentary formations correspond to a sequence of laterally-migrating continental ecosystems that overlapped according to their relative topographic position.**

Ice Age Climate Change

A stable Antediluvian crust prevented volcanic activity. The absence of volcanic ash in the upper atmosphere produced warm climates. Eruption of the "Fountains of the Great Deep" and the subsequent collapse of Earth's oceanic crust during the Worldwide Flood transferred heat from a 25 kilometer depth through the Antediluvian crust and into the atmosphere and oceans. Crustal compaction facilitated rapid conductive heat flow up through the post-Flood oceanic crust into seawater above, sustaining warm oceans and a warm atmosphere for centuries thereafter. Therefore, the Ice Age would not have commenced immediately following the Worldwide Flood Regression. *"Based on the cooling time for the ocean, the amount of time necessary to reach glacial maximum turned out to be a minimum of 174 years and a maximum of 1,765 years"* (Oard, 2005, p. 41). Most oceanic cooling took place during the Post-Flood Millennium.

"The ocean would have dissolved many nutrients and minerals from the sediments so that phytoplankton would grow rapidly in the warm, nutrient-rich conditions following the Flood. When phytoplankton is present, zooplankton can then grow rapidly to support the entire food chain. If phytoplankton bloomed worldwide for several hundred years after the Flood, it would probably have absorbed tremendous quantities of carbon dioxide, drawing down the concentration of carbon dioxide in the atmosphere. This, in turn, may have helped create a cooling of the atmosphere contributing to the "Ice Age" (Vardiman, 1997, p. ii).

The following accounts of volcanic eruptions causing atmospheric cooling provide small-scale examples of prolific volcanic activity, which was sustained during the Ice Age. The largest volcanic eruption of recent history was Mount Tambora, Indonesia, in 1815. Temperatures cooled worldwide and produced the "year without a summer" in 1816. *"Many crops failed to ripen and the poor harvest led to famine, disease, and social distress"* (Holt, 1996, p. 143).

The Rabul eruption on an island off New Guinea was probably responsible for the following descriptions in 536 AD. In Mesopotamia, *"the sun was dark and its darkness lasted for eighteen months; each day it shone for about four hours, and still this light was only a feeble shadow."* Meanwhile in Italy, Senator Cassiodorus wrote, *"The sun ... seems to have lost its wonted light, and appears of a bluish color. We marvel to see no shadows of our bodies at noon, to feel the mighty vigor of the sun's heat wasted into feebleness ... prolonged through almost a whole year ... a spring without mildness and a summer without heat"* (Holt, 1996, p. 140).

A similar dimming of the Sun in 44 BC is attributed to the explosive eruption of Mt. Etna, Sicily. Plutarch describes sunlight as follows: *"For during all that year its orb rose pale and without radiance, while the heat that came down from it was slight and ineffectual, so that the air in its circulation was dark and heavy, owing to the feebleness of the warmth that penetrated, and the fruits, imperfect and half ripe, withered away and shriveled up on account of the coldness of the atmosphere"* (Holt, 1996, p. 140).

The eruption of Toba *"resulted in temperature declines of up to 5 degrees centigrade and declines of 15 degrees centigrade at higher latitudes"* (Robock, et al., 2009). *"Not only do volcanoes produce large amounts of ash, which causes a short term cooling effect, but also sulfur dioxide (SO2), which can result in multi-year cooling effects by generating stratospheric aerosols"* (Gollmer, 2013, p. 2).

Sustaining an Ice Age for a century or longer required somewhat continuous volcanic activity. Glacial maximums were induced by periods of intense worldwide volcanic activity. Tropical to temperate climates persisted worldwide for approximately eight centuries following the Worldwide Flood as heat was progressively transferred from crust to oceans to atmosphere to space. **The Ice Age began as continental mountain upheaval dramatically accelerated and routinely supplied volcanic ash to the stratosphere (13-47 km) during the last two centuries of the Post-Flood Millennium.** Volcanic ash reaching the stratosphere requires a few years to fall back to Earth. Volcanic ash reflected a substantial portion of incoming sunlight back into space, causing atmospheric temperatures to cool worldwide. Temperature differential between warm oceans and a cool atmosphere increased markedly. Warm oceans "steamed" water vapor into the colder air above. Moisture-laden air rose, formed clouds, spread, and cooled over adjacent continents. Atmospheric temperatures decreased as increasing cloud cover reflected additional sunlight back into space. Abundant volcanic dust from mountain-building eruptions provided adhesion surfaces facilitating condensation of rain and snow.

Coastal lowlands became dreary and rain-drenched. Ice Age (Pleistocene ecosystems) were born as continental interior ice sheets began to *"accumulate rapidly closest to major storm tracks and the moisture source of the warm ocean. Glaciation would be delayed in areas that were too close to the warm ocean. The warm ocean heats the air above it and as this onshore flow of air spreads inland, it would keep the adjacent land relatively warm"* (Oard, 2004, Frozen in Time, p. 81).

Snow fell prolifically on cold continental interiors at high latitudes. Summer temperatures were too cool and cloudy to melt accumulated winter snow pack. Glaciers formed and reflected additional sunlight, making inland climates even colder. Continental glaciers expanded to hundreds of meters thick and blanketed large portions of continental interiors at high latitudes. The vast volume of ice contained in continental glaciers caused sea level to drop worldwide. Continental shelves and land bridges were exposed and soon became prolific grassland ecosystems.

Collapse Tectonics acknowledges only one Ice Age. Intermittent recessions of continental ice sheets occurred when mountain upheaval volcanic activity temporarily subsided, enhancing sunlight penetration to the surface and causing glaciers to temporarily retreat. Striated pavements, sometimes cited as evidence of pre-Ice Age (pre-Pleistocene) glaciations, are attributed to many processes, including submarine landslides (Oard, 1997, pp. 49-56), rapid water flow (Gore & Taylor, 2003, pp. 467-470), and ground-hugging volcanic eruptions (Perrett, 1935, pp. 76, 79).

Ice Age Extinctions

Cold weather extremes lagged behind pulses of mountain-building volcanic activity. Pleistocene ecosystem development lagged behind cold weather extremes. *"Throughout much of the world, most of the large mammals and many large birds became extinct at the end of the Ice Age ... Instead of being a time of cold, the* (Ice Age) *winters were actually mild ... Winters then become cold at the end of the Ice Age, much colder than today, especially at mid and high latitudes due to drier air, more sea ice, cooler oceans, and more reflection of sunlight from ice sheets. The animals were not adapted to cold winters and did not have time to adapt"* (Oard, 2008, pp. 17, 19).

Viking settlements located in what are now barren areas of Greenland imply major climate change during recent centuries. *"... the Zeno Map (1380) shows Greenland with no ice cap. The interior is filled with mountains. Rivers are shown entering the sea, in some cases at the points where at present great mountain glaciers are moving down through the mountains to the coast"* (Hapgood, 1996, p. 152). *"... seismic expeditions in recent years found that the under-ice topography agreed with the Zeno Map"* by verifying mountains halfway across a flat area in Greenland's interior (Hapgood, 1996, p. 152).

Mammoths thrived in dreary, rain-soaked grassland plains near modern coastlines and on areas now submerged as continental shelves. Mammoth tusks have been dredged from the bottom of the Arctic Ocean. After Arctic gales, the shores of islands are strewn with tusks. That implies much of the area between the islands and mainland was grassland where mammoths roamed during the Ice Age. Northern Siberia once provided more than half the world's supply of ivory from fossil mammoth tusks.

Glaciers acted as sediment traps for the accumulation of wind-blown silt. Retreating glaciers commonly deposited a remnant layer of loess (calcareous silt) on the ground surface. Wind erosion (eolation) fostered dust storms, which contributed to mammoth extinction (Oard, 2004, Frozen in Time, pp. 157-173).

The environmental backdrop above undoubtedly contributed to the reason why so many large mammals became extinct at the end of the Ice Age. However, abundant evidence of animals buried in icy permafrost tombs reveals another fascinating story of impending disaster and sudden permanent climate change.

The demise of Ice Age mammoths preserved in permafrost was quick. *"Microscopic examination of the skin showed red blood corpuscles, which was a proof not only of a sudden death, but that the death was due to suffocation either by gases or water ..."* (Whitley, 1910, p. 56). Vegetation found in the mouths and stomachs of frozen mammoths implies some mammoths were entombed during different seasons. The deaths of these mammoths apparently occurred during separate catastrophic events. Tree leaves and grasses that no longer grow in Northern Siberia are found in stomachs and between teeth of frozen mammoths. These plants are found today in the warmer Southern Siberian climate (Velikovsky, 1955, pp. 16-18).

The Tanana River Valley, Alaska, is covered by a frozen silt (or muck) that contains enormous numbers of trees and animals, including mammoth, mastodon, super-bison, and horse (Rainey, 1940, p. 305). Millions of animals were torn limb from limb and mingled with uprooted trees (Velikovsky, 1955, p. 13). Some frozen fragments retain ligaments, skin, hair, and flesh. At least four layers of volcanic ash may be traced in these deposits, although they are extremely warped and distorted (Hibben, 1943, p. 256).

Man-made artifacts similar to those recently used by Indians of the Tanana Valley, Alaska, were found under the tangled mass of bones and trees (Velikovsky, 1955, p. 280). Similar deposits are found on the western Yukon Peninsula, Koyukuk River, and Kuskokwim River. Gold mining excavations in the Fairbanks District of Alaska opened massive tombs of dismembered animals interspersed with splintered trees. Frozen skin, ligaments, hair, and flesh still cling to shattered bones of mammoths, mastodons, superbison, lions, and horses (Macgowan, 1950, p. 151).

On the south flank of New Siberia Island are "driftwood hills" 250 to 300 feet high. *"The trunks of trees lie flung upon one another in the wildest disorder, forced upright in spite of gravitation. The tops of tree trunks are broken and crushed as if they had been thrown with great violence from the south and heaped up on the bank ... Other hills on New Siberia Island and Koteloni are heaped up to an equal height with the skeletons of elephant, rhinoceros, bison, etc."* (Erman, 1848, pp. 376 & 383). *"... agents of a bituminous nature transformed them* (uprooted trees) *into charcoal, either before or after they were deposited and cemented in drifted masses of sand that became baked into sandstone"* (Velikovsky, 1955, p. 19).

Outburst Floods

The most probable cause of wood being altered to charcoal as interbedded sand is being baked into sandstone is the ground-hugging heat delivered by an air-borne *pyroclastic* flow. During the Ice Age, high elevation inland areas received more snowfall than was melted, causing continental glaciers to accumulate. Earth's crust isostatically sank beneath the overburden weight of thickening continental glaciers, unsealing vertical lineament fissures. Scalding *pyromagma* raced from depth to the surface, melting a vast quantity of glacial ice in hours. **Pyroclastic volcanic eruptions under glaciers generated outburst floods of rapidly flowing water, ice, and gravel surging out from under the base of a glacier.** Intermittent eruptions caused separate outburst floods, explaining mammoth entombment during different seasons.

"Outburst floods" at the end of the Ice Age launched surge waves down river valleys, eroding sediments as they raged onto coastal plains. Surge waves of water, ice, pyroclastic ash, and silt smashed and entombed mammoths and other large mammals as they grazed along river valleys. Grassland animals were drowned, buried, and quick frozen in icy, sediment-laden tombs.

The immense volume of water suddenly released by a large outburst flood inundated the grassland coastal plain. Animals fled the icy waters and congregated together on the highest ground as rising floodwaters spread across distant grasslands. As the floodwaters rose above hilltops, the animals trampled each other in a desperate attempt to climb to higher ground and avoid drowning in the rising frigid water.

Volcanic ash expelled into the atmosphere by outburst eruptions potentially explains a sudden climate chill, preserving large animals in permafrost. Volcanic ash and sediments from those same sub-glacial eruptions may have impeded circulating warm water currents flowing into the Arctic Ocean from the Pacific and/or Atlantic Oceans. Reduced circulation from warmer ocean currents may have caused the Arctic Ocean to freeze over during winters, producing a colder drier climate and sustaining permafrost conditions. That also potentially explains the abrupt change in vegetation.

Water from snow and ice melted by a sub-glacial volcanic eruption generated a catastrophic "outburst flood" on November 5, 1996 in Iceland. The peak discharge of 45,000 cubic meters per second rapidly deposited extensive rhythmic sediment accumulations during the 36 hour flood (Russell & Knudsen, 1999, pp. 1-10; Snelling, 1999, pp. 46-48). *"Landforms indicative of cataclysmic outburst floods ... include flood-scoured channelways, giant bars, and gravel wave trains"* (Baker, et al., 1993, p. 348).

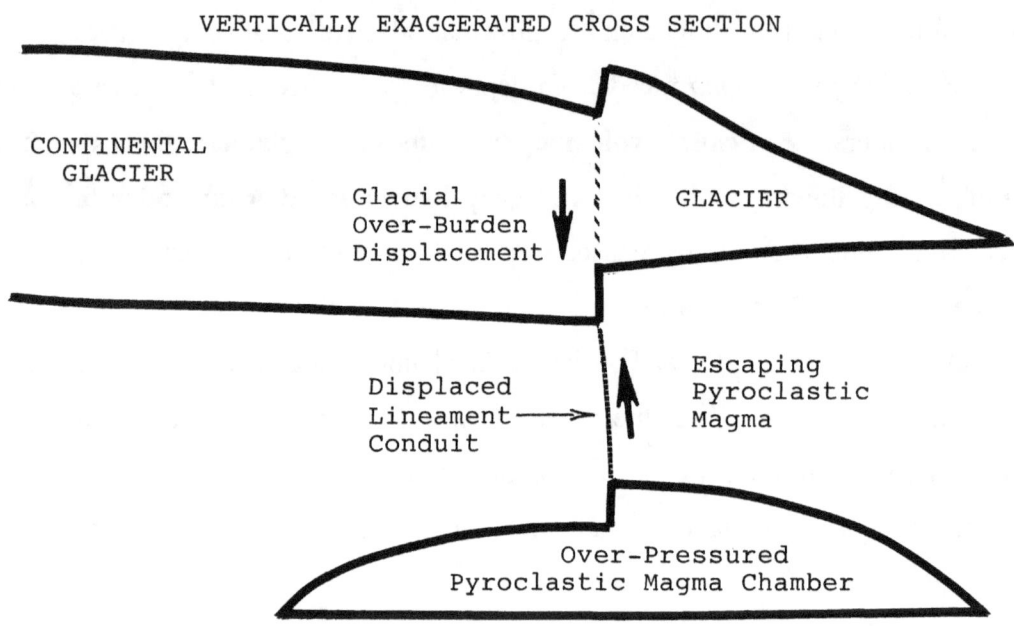

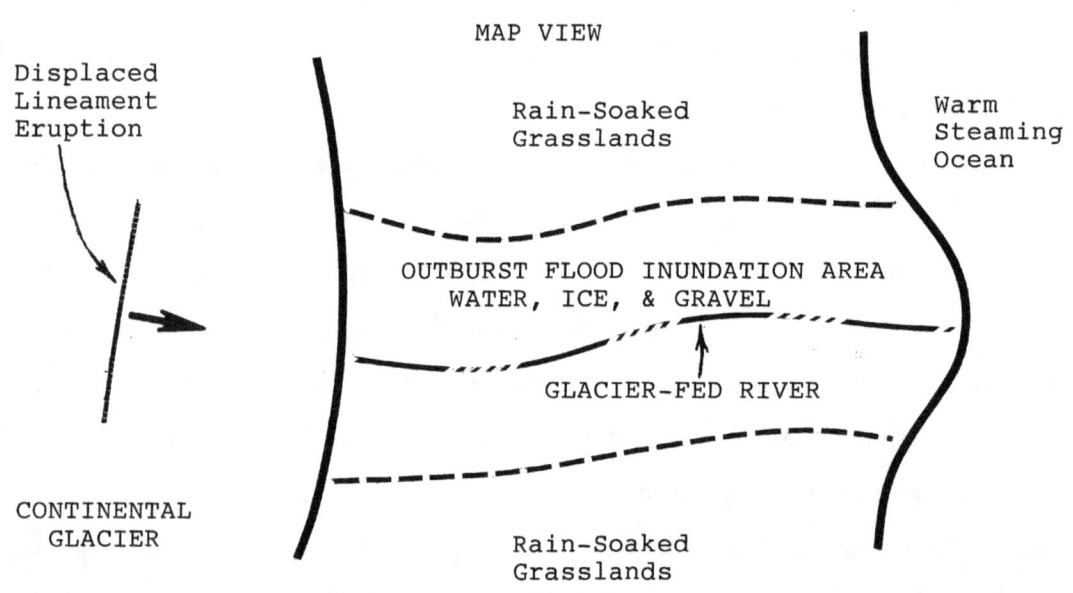

The Liakhov Islands in the Arctic Ocean contain abundant mammoth remains. *"Such was the enormous quantity of mammoths' remains that it seemed ... that the island was actually composed of the bones and tusks of elephants, cemented together by icy sand"* (Whitley, 1910, p. 41). The Arctic Ocean's New Siberian Islands also contain abundant bones. *"The soil of these desolate islands is absolutely packed full of bones of elephants and rhinoceroses in astonishing numbers"* (Whitley, 1910, p. 36).

A similar flight to high ground is also evidenced further onshore. *"The best Mammoth bones, as well as the greatest number, are found at a certain depth below the surface usually in clayhills more rarely in black earth. The more solid the clay, the better the bones are preserved. Experience has shown that more are found in elevations situated near high hills than along the low coast of the flat tundra."* (Howorth, 1880, p. 181). These mammoths were grazing in locations allowing flight to higher ground when the outburst flood struck. However, the avoidance of icy gravel entombment from the outburst flood was immediately followed by pyroclastic ash falling from the sky and death by suffocation. Thick deposits of pyroclastic ash rapidly covered their bodies. The volcanic ash rapidly oxidized and decomposed into clays. Minerals released by blanketing clays partially fossilized and helped preserve these mammoth bones.

Dinosaurs in Permafrost

The Liscomb Bone Bed along the Colville River of the North Slope of Alaska contains frozen, un-fossilized dinosaur bones. Tough tail ligaments are still attached to some dinosaur tails (Creation Magazine, 1990, pp. 10-14; Buddy Davis, pers. comm., 1998). Ligaments deteriorate rapidly in an unfrozen state. **According to the Geologic Time Scale, the dinosaurs became extinct approximately sixty-four million years before the Ice Age began. However, the Ice Age catastrophe that apparently killed, buried, and preserved these dinosaurs in permafrost took place no more than a few thousands of years ago!**

Antarctic Ice Cap Formation

The discovery of thousands of well-preserved leaves in Antarctica has sparked a debate among geologists, concerning when the polar ice cap formed. *"The leaves compressed by subsequent layers of ice, look like fossils. But unlike fossils ... the leaves retain their original cellular structure and organic content"* (Raymond, 1991, p. A11). Because this delicate organic matter would gradually deteriorate during prolonged freezing, the quality of leaf preservation implies the Antarctic ice cap formed rapidly only a few thousand years ago.

The Orontenus Finaeus map drawn in 1531 shows *"mountain ranges that skirted the coasts suggested the numerous ranges that have been discovered in Antarctica in recent years ... The mountain ranges were individualized, some definitely coastal and some not. From most of them rivers were shown flowing into the sea, following in every case what looked like very natural and convincing drainage patterns. This suggested, of course, that the coasts may have been ice-free when the original map was drawn"* (Hapgood, 1996, pp. 79-83). **The ancient seafarers who drew the original maps from which the Orontenus Finaeus map was eventually copied probably sailed around the world prior to the maximum extent of Ice Age glaciation.** This implies a seafaring civilization may date back through the Post-Flood Millennium to some of Noah's grandchildren.

Ice Age Ends

"Once a snow cover is established, the temperature cools about another 6 degrees Centigrade because of snow's greater reflectivity to sunlight. This reinforces the cooling already caused by volcanic ash and gases" (Oard, 2004, Frozen in Time, p. 79). However, snow-reflection cooling can only be sustained long term if continuing volcanic ash contributions to the upper atmosphere offset the ash falling to Earth. The completion of continental basin collapse and corresponding mountain upheaval volcanics eventually allowed atmospheric dust to clear, atmospheric temperatures to rise, ocean evaporation to subside, snowfall to decrease, and Ice Age glaciers to recede. During the Ice Age, Earth's oceans were progressively cooling from excessive evaporation and snowmelt contributions.

The contrast between warm oceans and cool atmosphere dissipated, producing warmer, drier climates. Ocean levels rose, inundating shallow continental shelf areas.

The Ice Age signaled the end of the tectonically-active Post-Flood Millennium. Earth is a tectonically-active planet. The transition from frequent catastrophic processes to slow and gradual processes explains why it is inappropriate to extrapolate modern geomorphic rates back through time for interpretation purposes.

POST-FLOOD DILUVIAL FEATURES

"Water gaps", "planation surfaces", "sheetflow conglomerates", and "pediments" are erosion and deposition features that were produced by hydrodynamic flows on a larger scale than those observed in the modern world. *"Where joints and faults were present following the Flood, they would have formed natural river courses. In some cases these features might help explain how rivers cut through mountains and topographic highs, or have cut exceptionally deep canyons in short periods of time ... Not only would large amounts of post-Flood erosion be expected, but thick deposits of the eroded material would follow ... In all likelihood, these processes would have totally removed any planation surfaces, water gaps, pediments and other such features that have been imagined forming just before the continents were lifted out of the oceans at the Flood's end. In my opinion, it is unthinkable that these kinds of features would have survived the post-Flood world unless they were immediately buried"* (Whitmore, 2013, pp. 10, 13, 18). However, these features exhibit no evidence of subsequent burial.

Porous crust collapse generated lateral crustal displacement, uplifting adjacent mountain ranges. Continental root underplating tilted areas the size of several states. Shallow continental interior seas were tilted, causing dam breaches that suddenly spilled sheetflows across large regional areas. **Diluvial features are most reasonably explained by post-Flood processes.**

Water Gaps

Water gaps are remnant erosional features, which are present across topographic highs in directions unrelated to modern stream and river flow patterns. Modern streams and rivers form after the land is subaerially exposed, following longer paths around topographic barriers. Topographic "saddles" commonly bisect ridgelines where fault displacement has shattered bedrock. **Accelerated erosion along the weakened fault line produces topographic "saddle" features that may also be described as "water gaps".**

Planation Surfaces

Planation is a form of erosion which carves flat surfaces. All continents exhibit large topographically flat areas called planation surfaces. Examples include the plains of Africa and Australia. There is no modern analog of a planation surface forming today. Planation surfaces are now being dissected by erosion. *"Erosional processes today do not produce the flat landscapes that were produced in the past. Modern day erosional processes roughen surfaces, forming rills, coulees and valleys. Today, we observe that previously-planed surfaces are dissected"* (Oard, 2002, (16)3, p. 40).

Planation produced flat surfaces, without regard to differing hardness of individual layers eroded (Paige, S., 1912, p. 444; Johnson, D., 1931, p. 174; Small, R.J., 1978, p. 319). This implies that sheetflow erosion occurred prior to consolidation (hardening) of the rock, and/or the erosional mechanism was unusually powerful. The flatness of planation surfaces is enhanced by a thin veneer of coarse gravel, which sometimes contains cobbles and boulders (Twidale, C.R., 1981, p. 425; Thomas, M.F., 1994, p. 245). The veneer materials are typically rounded or sub-rounded, implying water transport. **Planation surfaces were eroded by post-Flood dam breach sheetflows, which suddenly released large volumes of water from "perched" continental interior seas.** Broad areas were flattened by the sudden dam breach deluge.

Sheetflow Conglomerates

Conglomerates are sedimentary rock bodies consisting of boulders, cobbles, pebbles, sand, silt, and clay. Transportation of conglomerates from source area to deposition area depends upon water velocity, depth, and slope. For example, fine-grained sediments are transported more easily than coarse-grained sediments. Transport of disc-shaped boulders requires faster water velocity and/or steeper slopes than rounded boulders of equal volume.

Modern conglomerates are deposited in channel-cut river bottoms during raging floods. In contrast, some pre-historic conglomerate formations blanket hundreds of thousands of square miles. *"The Shinarump Conglomerate and recognized equivalents outcrop over wide areas in northern Arizona and southern Utah and less extensively in central, northern and western Utah, western Colorado, northwestern New Mexico, southeastern Nevada, and southwestern Idaho. Its total original aerial extent exceeded 125,000 square miles. It is usually less than 50 feet thick but may locally reach 300 feet"* (Stokes, 1950, p. 91).

The Shinarump Conglomerate is continental and fluvial (water-deposited) in origin as indicated by primary structure, fresh-water mollusks, invertebrate fossils, petrified wood, broken bone fragments, and paleo-geographic position. The Shinarump Conglomerate is amazingly thin for a continental deposit blanketing a multi-state area. Some of the rocks have traveled hundreds of miles to their present location, where they are mixed with fragments from underlying beds (Stokes, 1950, pp. 91-97).

A distinctive erosional unconformity marks the base of the Shinarump, with localized channels cut into underlying sediments. The bedrock surface below the unconformity usually appears fresh with little or no weathering. However, groundwater flowing through the base of the conglomerate may leach the underlying bedrock, causing the false appearance of weathering (Stokes, 1950, p. 96). Soil zones below the basal Shinarump unconformity are conspicuously absent, indicating the soils eroded or did not have time to form. The evidence implies the Shinarump Conglomerate was transported as a massive sheetwash across a broad, gently sloping erosional surface.

At nearly all locations, the Shinarump Conglomerate grades upward into the overlying beds. That implies a rapid initial deposition event dissipating upward into subsequently slower deposition processes. This "rapid-to-slow" depositional sequence was mimicked at later times during deposition of the Buckhorn and Dakota Formations (Stokes, 1950, pp. 92-93). **These widespread sheetflow conglomerates are attributed to dam breaches of continental interior seas.**

Pediments

Pediments are planation surfaces that were tilted in association with adjacent mountain uplift as sheetflow subsided and deposited its basal load of veneer gravels. Pediments are herein defined as large planar surfaces typically tilting a few degrees from horizontal away from an adjacent mountain range. *"Pediments are not observed to be forming today"* (Oard, TJ, 2004, (18)2, p. 15). Pediments are *"reported on six continents. Their distribution spans the range of subpolar latitudes from the Arctic to the Antarctic and the range of climate from hyperarid to humid tropical"* (Dohrenwend, 1994, p. 321). So pediments appear to be a hydrologic feature.

Pediments have not formed since rapid mountain upheaval subsided at the end of the post-Flood Millennium. A thin veneer of rounded to sub-rounded gravels covers some pediment surfaces. This is typical of sheetflow deposition rather than stream flow deposition, which is currently forming channels and dissecting pediments today. *"A minor portion of the coarse gravel capping some pediments is exotic (foreign) to the mountains adjacent to the planation surface"* (Crickmay, C.H., 1975, p. 108). These foreign materials indicate sheetflow transport originated from a distant source.

If the coarse gravel veneer had been sourced exclusively by adjacent uplifting mountains, the veneer would thicken towards the adjacent mountains. Flow velocities transporting those gravels down steep mountain slopes would have immediately slowed and deposited their sediment load at the base of the mountain slope. In contrast, *"the coarse-gravel veneer normally thickens away from the mountains"* (Oard, 2004, TJ 18(2), p. 16).

The typical linear, sharp, angular, contact between pediments and mountain slopes, begs for a depositional source other than the adjacent mountain. We sometimes observe modern alluvial fans at the base of mountain slopes, which produce intermittent convex upward cones at the mountain edge. However, where alluvial cones exist, they appear to be comparatively tiny modern depositional features superimposed upon a much larger host pediment surface. In contrast, pediments typically exhibit a slight concave upward profile, steepening slightly towards the adjacent mountains (Hadley, 1967, pp. 83-89; Mammerickx, J., 1964, pp. 417-435). **Apparently, adjacent mountains were being uplifted and pediment surfaces were being tilted as sheetflow running parallel to the mountain flank was subsiding and gravels were being deposited.**

"The top of the pediment on one side of the mountain range can be higher than the one on the other side!" (Oard, 2004, TJ 18(2), p. 19). This may have been caused by different sheetflow water levesl at time of deposition, by separate sheetflow events, or by differential pediment uplift following deposition.

Grand Canyon Stratigraphy

The Grand Canyon of Northwestern Arizona, U.S.A. is a magnificent chasm which provides westward drainage from four merging rivers - the Colorado, San Juan, Green, and Little Colorado – of Arizona, Utah, Colorado, and New Mexico. Mysteriously, this large drainage area is situated at a lower elevation than the Colorado Plateau, through which the Grand Canyon portion of the Colorado River flows.

Igneous and metamorphic basement bedrock consists of Antediluvian Vishnu Schist intruded by Zoraster Granite. Both were erosionally truncated to form the Greatest Angular Unconformity during the Transgression Stage of the Worldwide Flood. Nine sedimentary formations comprising the Proterozoic Grand Canyon Supergroup were rapidly deposited during the Worldwide Flood as the area subsided. Grand Canyon Supergroup sedimentary formations were tilted and eroded by rapid sheet flow during the Worldwide Flood Regression. Regression Stage erosion subsequently carved the Great Unconformity, forming a flat peneplain surface interrupted by scattered monadnocks.

The earliest post-Flood deposition regime in the Grand Canyon area corresponds to the Cambrian Tapeats Sandstone / Bright Angel Shale / Muav Limestone marine transgression sequence as tectonic subsidence continued. The gently-sloping, denuded land surface experienced rapid erosion, surface flow, and westward transport of materials into the adjacent shallow continental interior sea. Sediments were transported by bottom-hugging flows in early post-Flood rivers and streams. Upon reaching the edge of the continental interior sea to the east, water velocity slowed, causing sand-sized materials to immediately settle to the bottom. Wave action spread the sands along the coastline, winnowing out clay-sized particles.

Clay and silt-sized particles remain in suspension much longer than sands. For example, *"particles of these* (clay and silt) *sizes would take 3.2 years and 120 days respectively to fall 100 m in still water"* (Allen, David, 1996, p. 362). Although shallow sea depths were less than 10 meters, wave action, tidal motion, river inflow, and river outflow prolonged suspension and extended lateral distribution prior to deposition. Time of suspension allowed silts and clays to be laterally transported offshore for tens of kilometers before settling to the seafloor.

Further offshore, carbonate muds were intermittently extruded into seawater from vents sourced by collapsing porous crust below. Those carbonate muds spread outward from vents and settled slowly to the bottom in the form of micritic (1-4 micrometer grain size) deposits that subsequently hardened into limestones and dolomites. (Micrite carbonates are contrasted to larger grain-sized carbonate materials precipitated by marine life and deposited as shell fragments.)

Meanwhile, low-density *hypomagma* rose in continental roots, butted into the crustal ceiling and spread outward, progressively underplating the area to the west. Consequent isostatic uplift generated a slight eastward tilt amounting to a tiny fraction of a degree on the overlying crust. That tilt progression periodically extended marine indundation eastward several kilometers at a time. Deposition of shoreline Tapeats sandstone, nearshore Bright Angel Shale, and offshore Muav Limestone intermittently shifted eastward as the shallow sea transgressed across the land, forming distinct contact surfaces between those characteristic lithologies. At any specific time, that progressive eastward marine transgression caused Bright Angel Shale to overlap Tapeats Sandstone, followed by Mauv Limestone overlapping Bright Angel Shale.

Intermittent porous crust collapse events formed sinking sediment traps on the shallow seafloor above. Shallow continental interior seas routinely transgressed and regressed across the area depositing a series of Paleozoic and Triassic sedimentary formations. Sediments were sourced by volatile extrusives expelled from the collapsing porous crust below and by clastic materials eroded and transported from adjacent subaerially-exposed areas.

At any specific time, shallow marine, intertidal, deltaic, fluvial, mudflat, sabkha, and eolian dune ecosystems were all actively being deposited at various locations across this sinking regional sediment trap. Those ecosystems repeatedly transgressed and regressed across the top of adjacent sediments. Ripple marks, raindrop impressions, mud cracks, worm burrows, plant roots, crawling traces, feeding marks, and animal footprint trails, are exhibited within these laterally-shifting ecosystem habitats. These distinctive habitats were stacked one above another as sedimentary layers accumulated to a depth of three kilometers. Fossils and geologic context affirms that these ecosystem habitats were locally-established and deposited at their post-Flood habitat location.

A continental root was continuing to source *hypomagma* underplating Earth's crust, causing gradual, isostatic uplift. Porous crust collapse beneath the future Colorado Plateau had subsided by the end of the Triassic. Isostatic uplift of the Colorado Plateau began to outpace collapse-induced subsidence. Eroded materials from distant sources no longer blanketed the area with new sedimentary formations. Shallow inland seas shrank into lakes and dried up. One large river meandered across the gently dipping plain.

Isostatic uplift caused by continental root underplating did not occur uniformly. Instead, uplift produced a subtle rotating tilt of the rising Colorado Plateau from dipping northeast then east then southeast then south then southwest then west. Surface drainage systems shifted accordingly. McKee, et al. (1967) propose that the ancestral Colorado River initially followed its present course southward to the eastern end of what would become the Grand Canyon. However, instead of turning west along its modern course, the ancestral Colorado River sluggishly flowed southeast along the course of the modern Little Colorado River and Rio Grande River to the Gulf of Mexico.

Continuing porous crust collapse of Great Plains to the east generated westerly displacement, beginning to buckle up the Rocky Mountain chain to the east of the Colorado Plateau. Meanwhile, the ancient Lake Bonneville inundated a topographic low area east of the rising Colorado Plateau and west of the uplifting Rocky Mountains. Aerial photographs reveal segments of the extended Lake Bonneville paleo-shoreline footprint as stretching far beyond the modern Lake Powell shoreline. Intermittent Rocky Mountain upheaval events raised the Lake Bonneville area and tilted the entire region, including the Colorado Plateau surface, westward.

"Drainage systems evolve continually – and do so chiefly through headward erosion and capture and in response to tectonic movements" (Lucchitta, 1990, p. 316). **Stream piracy is achieved by headward erosion, which intercepts and captures stream flow from an adjacent drainage system.** During the latter part of the Post-Flood Millennium, a youthful, steep-gradient stream, which emptied into the Gulf of California, was rapidly extending its drainage eastward across the Colorado Plateau through unconsolidated or weakly consolidated near-surface sediments. Each time Rocky Mountain upheaval tilted the Colorado Plateau westward, headward erosion of the stream extended eastward, by deeply eroding along the ancient meandering river path. Large lakes near that advancing stream were periodically intercepted by headward erosion and catastrophically drained. That rapid, steep-gradient flow carved dead-end side canyons down into the deeply-incised, westward-flowing stream.

Headward erosion by that steep-gradient stream eventually crossed the Colorado Plateau from west to east, breaching the ancestral Colorado River. A torrent of water was suddenly diverted westward into the steep-gradient stream, rapidly eroding and deepening the Grand Canyon channel. Headward erosion and channel deepening extended northward up the Colorado River Valley and breached ancient Lake Bonneville. Another torrent of water rushed down the Colorado River, eroding and deepening the Grand Canyon channel.

Headward erosion and channel deepening proceeded more slowly to the southeast, reversing flow direction along the Little Colorado River channel. Eventually, headward erosion breached the ancient Lake Bidahochi, located in the Little Colorado River Valley (Scarborough, 1989). Another torrent of water from the ancient Lake Bidahochi rushed northwest and turned west into the Grand Canyon, eroding and further deepening the chasm. The Colorado River drainage system was finally established in its modern position. Subsequent isostatic uplift of the Colorado Plateau produced topographic barriers to flow, which were quickly removed by erosion of the rushing Colorado River.

A vast volume of rock formerly occupied the area which was vacated by erosion of the Grand Canyon. Sediment overburden removal caused isostatic uplift, elevating the Grand Canyon rims and valley. That uplift induced additional erosion of the Grand Canyon valley floor.

ECONOMIC APPLICATIONS

Ore Mineral Intrusion

Long, narrow mineral belts produce most of the world's ore minerals. Bedrock surrounding highly acidic precious metal intrusions frequently exhibits hydrothermal alteration. The near-surface oxidation of pyrite stains bedrock yellow, orange, and red. Prospectors have scoured mountain regions, digging test pits in areas exhibiting colorful hydrothermal alteration. **Mineral belts are typically located along the surface exposure of uplifted lineaments, which provided conduits for ore mineral intrusion associated with mountain upheaval overpressuring at depth.**

Melting points of metallic compounds contained in *pyromagma* are low. However, their specific gravity is too high to float up to the base of the downward-solidifying crust. During the latter stages of continental basin collapse, remnant *pyromagma* at the base of the crust contained high concentrations of free-flowing metallic sulfides. Late Post-Flood Millennium collapse events in adjacent continental basins generated compression-driven flow of mineral-rich pyromagmas laterally below adjacent uplifting mountain ranges. Those mineral-rich pyromagmas were forced into thrust faulted lineaments and solidified in the form of narrow mineral belts.

Rock-forming components of displaced *pyromagma* solidified first, due to higher solidification temperatures of those minerals. As metallic sulfide-rich pyromagma rose near the surface, it contacted circulating groundwater and atmospheric oxygen. Solidification segregation of metallic compounds from rising pyromagma was determined by temperature, viscosity, pressure, oxidation/reduction, pH, and the specific gravity of each metal. For example, as a gold vein is excavated deeper, the concentration of silver and iron compounds usually increases, with a corresponding decrease in gold.

Precious metals are commonly associated with pegmatite dikes because both have low melting points, causing them to solidify under cooler temperatures close to the surface. Repeated tectonic compression during upheaval events fractured pegmatite dikes, periodically opening conduits for intrusion and deposition of precious metals-rich pyromagma.

PRECIOUS METALS ORE EXPLORATION

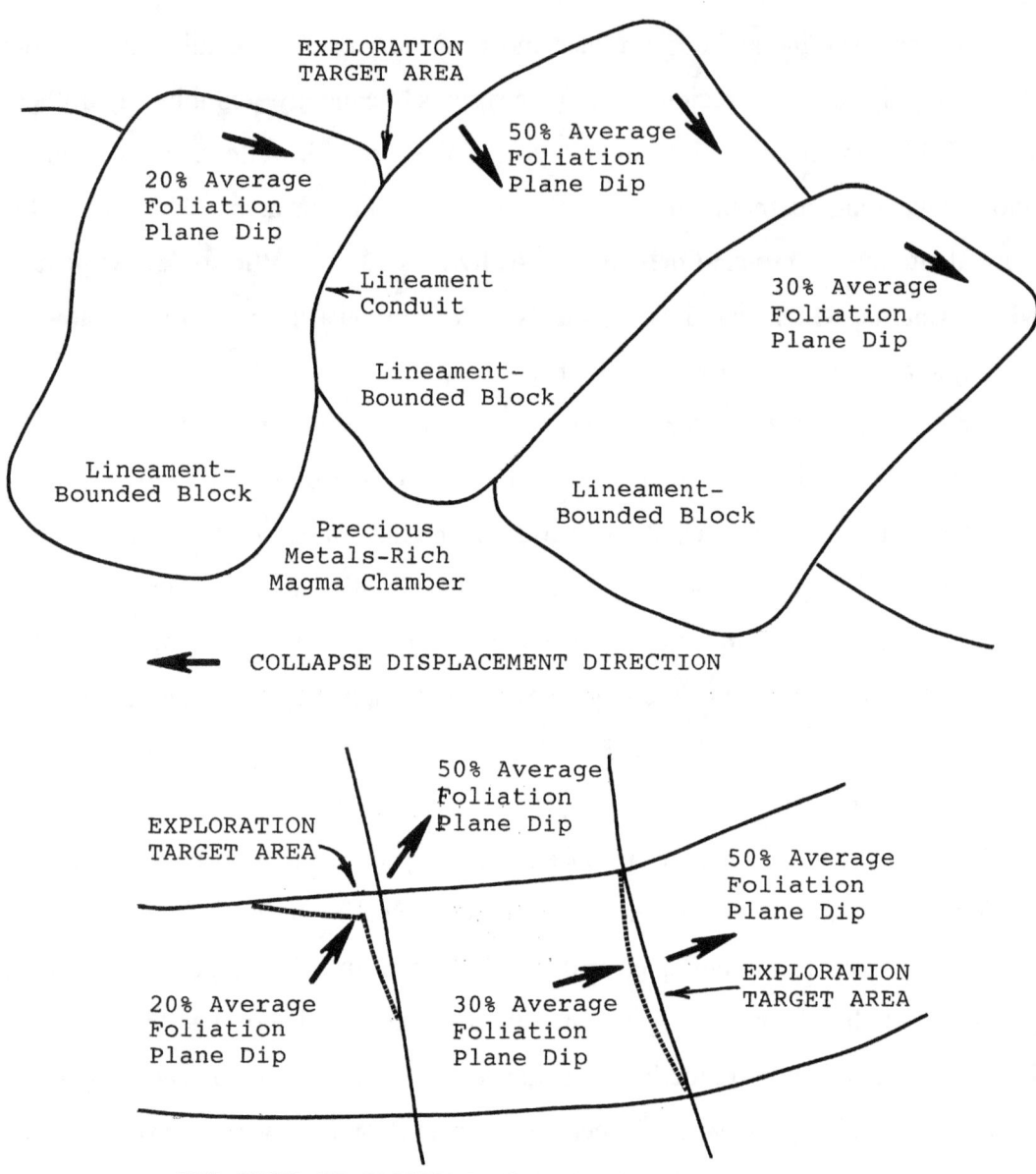

Placer Deposition

Placers are the "ore of sedimentation". Minerals such as gold are eroded, transported, and concentrated in alluvial (water-laid) deposits. Mysteriously, *"most placer deposits of economic value are in rocks of Tertiary age or younger..."* (Unknown, 1974, p. 48). Precious metals-rich remnant *pyromagma* commonly migrated close to the surface before solidifying. Mountain upheaval and tilting steepened surface slope angles, generating rapid erosion. Those eroded sediments sometimes contained high concentrations of precious metals. Gold and other high-density minerals concentrated in gravelly conglomerates, where rapid flow velocities slowed. **Most mineral-rich placer deposits are found in Tertiary and younger sediments, due to the late timing of mountain upheaval, mineral injection, and subsequent rapid erosion.**

Coal

The surface of tsunami backwash sheetflow moved more rapidly than the base of the sheetflow due to resistance caused by water flowing across stationary ground. Therefore, vegetation floating in a receding tsunami sheetwash moved faster and further than other constituents from each ecosystem. Backwash sheetflow carried abundant floating vegetation back to the edge of the continental basin sea, where flow velocity quickly slowed and vegetation settled out of suspension. **Paleoshoreline deposits contain the highest concentration of vegetation.** Small boulders were often ensnared by roots of uprooted trees that piled up along shorelines.

Reducing conditions along continental basin shorelines were generated by hydrogen sulfide expelled during each collapse event. Hydrogen sulfide gas is heavier than air and temporarily hovered over the water and along ancient shorelines. Reducing conditions were generated during and immediately following each collapse-induced tsunami event. Mildly-acidic sea water inhibited bacterial decomposition of recently-deposited organic material.

Rounded boulders in sheetflow waves tumbled down slope in retreating tsunami backwash waves. However, when a boulder rolled onto a thick layer of loose vegetation, it sank and came to rest. Boulders encased in coal seams commonly endanger miners by falling from tunnel ceilings. Rapidly buried vegetation layers eventually compressed to approximately one-tenth of their original thickness, due to dewatering and overburden compression from later sediment deposition above.

Cyclothems often contain pyrite (iron sulfide) in organic-rich lenses such as coal. Exposure to oxygen at the surface prevents pyrite formation. The lack of oxidation in coal lenses confirms rapid burial under reducing conditions. Subsurface flow of groundwater recharged during a recent tsunami inundation temporarily sustained acidic groundwater conditions. Freshly-buried paleoshoreline vegetation altered into coal. That explains why Carboniferous (Mississippian/Pennsylvanian) paleoshoreline formations commonly contain rich coal deposits.

Petroleum

Organic materials begin to decompose immediately after burial, as observed by methane release from landfills. Biogenic hydrocarbons have been discovered in recent onshore, nearshore, and offshore sediments along the coast of Texas and Louisiana and in the Gulf of Mexico (Smith, P.V. Jr., 1952, pp. 437-439).

Petroleum formation, migration, and entrapment follow a routine progression. Plants and animals were rapidly buried, preventing rapid surface decomposition. Continuing sedimentation increases depth of burial. Loose sediments lithify (harden) into bedrock as sediments dewater, overburden pressures increase, and temperatures rise. Biogenic materials progressively brake down, forming oil and natural gas.

Salt water commonly fills intergranular porosity within sedimentary reservoir rock. Most petroleum components are of lower density than salt water. Therefore, low-viscosity forms of oil and gas displace salt water above as it rises between mineral grains in permeable reservoir rock. However, overlying sedimentary layers known as "cap rocks" are too impermeable to transmit most gases and liquids. Rising oil and gas bumps into the cap rock barrier, flattens against that tilted ceiling, and migrates laterally up-dip. If the cap rock forms a closed high or the reservoir rock becomes impermeable in the updip direction, the petroleum is trapped in a "pool" within bedrock porosity. When the pool is penetrated by a drill hole, the oil flows into the hole, is pumped to the surface, and transported to a refinery.

"*The organic theory of petroleum now prevails ... petroleum was formed from dead organisms, long buried in the crust, and converted by heat and pressure into hydrocarbons*" (Lalomov, 2007, p. 65). Biogenic materials preserved under favorable conditions (pressure, heat, time, chemical catalysts, and preservatives) do generate oil and natural gas. However, simple hydrocarbons are also released during volcanic eruptions. Therefore, it is foolish to ignore non-biogenic hydrocarbons rising from Earth's interior as a potential contributing source for petroleum reservoirs.

Natural processes do not generate perfect seals, allowing all petroleum reservoirs to leak. When production of an oil field becomes uneconomical, depleted wells are plugged. Non-producing fields are occasionally monitored in the hope that slowly migrating oil will partially replenish the field and higher future oil prices will reestablish economic viability. However, measurements typically record a progressive decline in reservoir pressure, indicating these plugged depleted fields are continually leaking.

Based on the principle that native petroleum reservoirs leak, one Denver area service company set gas collection canisters in soils on prospect drill sites to confirm the existence of a petroleum reservoir below. The petroleum reservoirs I was prospecting for were supposedly deposited between 300 million and 700 million years ago. The problem is that petroleum reservoirs leaking at a rate fast enough to capture and measure in soil canisters, after brief (usually one day) capture times, would be expected to deplete within several thousand years. Canister sampling of reservoirs hundreds of millions of years old would be useless. However, if petroleum reservoirs are only a few thousand years old, that soil gas sampling technique would be a useful confirmation tool.

"One problem with this model is the persistence of elevated reservoir pressures over long periods of time. In real conditions it is difficult to imagine sedimentary rocks remaining so impermeable for so long. Given millions of years, one would expect these pressures to equilibrate in the subsurface, even in rocks of very low permeability. In natural conditions, oil and gas usually are under high pressures (the major lifting mechanism to transport oil and gas through wells to the surface), and this may rightly be considered an argument for the young age of these deposits" (Lalomov, 2007, p. 65). If oil and gas reservoirs were millions of years old, the low-density gas fraction of petroleum would have long since depleted. Reservoirs would not be sufficiently overpressured to dramatically expell oil gushers when penetrated by the drill bit.

A second problem with petroleum age is the presence of radioactive carbon-14 in oil and gas. The carbon-14 isotope is assumed to have been absorbed by living plants and animals millions of years earlier. However, with a radioactive half-life of approximately

5,735 years, the original carbon-14 isotope would have decayed entirely, and should be absent from the oil and gas reservoirs (Barenbaum, 2004; Baumgardner, et al., 2003; Doughty, 2006). In this case, groundwater supplementation of carbon-14 from surface sources would not occur because circulation would allow the petroleum to escape the trap. Without a recent contributing source for the carbon-14 isotope, oil and gas reservoirs are presumed to be no more than several thousand years old.

A third problem is commonplace examples of reservoir replenishment occurring on the scale of decades, not millions of years (Dmitrievsky and Valyaev, 2002). Source rock millions of years old could not continue to generate oil and gas at a rapid rate. However, if the source rock is only a few thousand years old, significant oil and gas generation and accumulation would still be continuing.

Today, we do not observe the massive rapid burial processes necessary to reasonably source the vast number of oil and natural gas fields. Deposition environment modeling has become an extremely effective tool in petroleum prospecting. Petroleum geologists map sedimentary layers and sequence tectonic movement to define drilling prospects. Each sedimentary formation has unique characteristics that reveal where exploration prospects are most likely to discover new oil and gas fields. Geologic evidences provide a wealth of information regarding sequence, but almost nothing about geologic time. **Source rock deposition, petroleum generation, migration, and entrapment favor a comprehensive time-frame spanning thousands (not hundreds of millions) of years.**

Carbonate Exploration

Continental basin seas precipitated carbonate reservoirs, such as the Ordovician Red River Formation of the Williston Basin in Montana and North Dakota, U.S.A. Collapse events produced greater outward displacement at depth, rather than near the surface. That tilted lineament-bounded seafloor blocks back toward the collapsing continental basin center. Each tilted block produced a paleo-topographic high above the corner or edge of the block closest to the continental basin rim.

PETROLEUM EXPLORATION - CARBONATE BIOHERMS

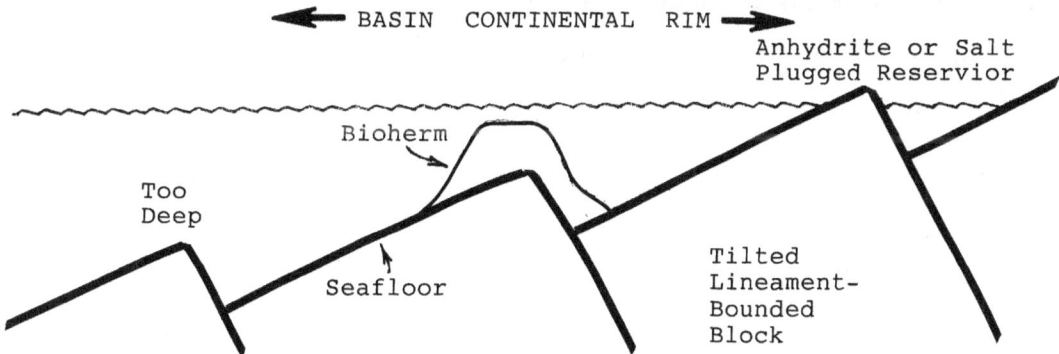

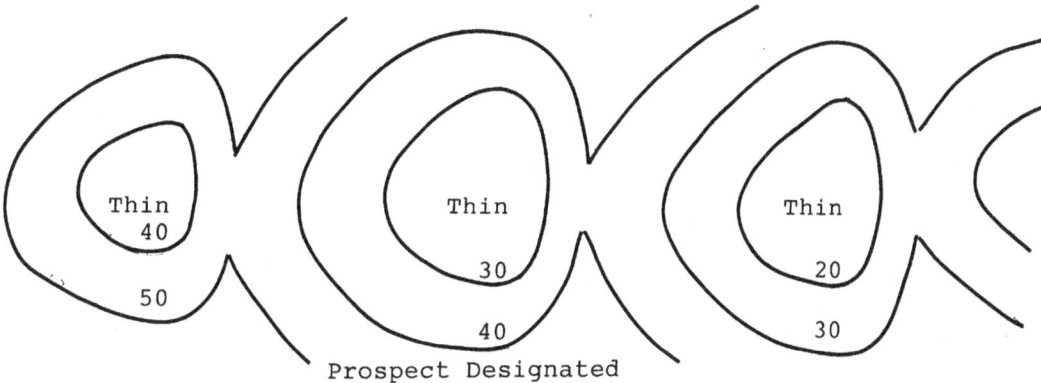

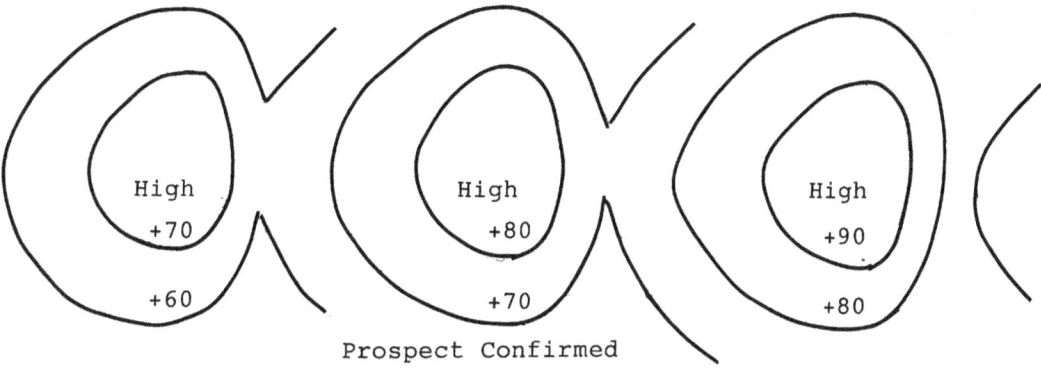

Sunlight penetrated shallow clear waters to facilitate prolific plant and animal life. Organic mounds called "bioherms" were built by sedentary (stationary attached) organisms including coral, algae, and stromatoporoids. **Bioherms formed at shallow depths on isolated offshore paleo-topographic highs at up-thrown lineament block corners and edges.** Bioherm organisms require clear water, so nearshore waters muddied by wave action were not suitable. Associated burrowing, shell debris, and plants increase organic content and void space in entombed bioherms. Those void spaces provide reservoir storage capacity for organic material decomposing into petroleum.

Short interval isopach thins (or geophysical isochron thins) above the reservoir rock level are the primary target. An isopach thin in the sediments above the reservoir indicates a paleo-topographic seafloor high existed before being subsequently blanketed by accumulating sediments. Petroleum, being of lower density than water, migrates upward through permeable rock and collects beneath impermeable rocks above. Structural highs (or geophysical time structure highs) at reservoir depth are used to confirm the prospect.

Anhydrite or salt plugging of the target reservoir indicates the zone was intermittently emerged before burial, and is not a suitable reservoir. Explore for similar prospect features closer to the basin center. In contrast, poor reservoir quality in limestones and dolomites imply the feature was too deep to support a shallow photic zone bioherm. Explore for similar prospect features further away from the basin center.

Thick shallow-formation isopach intervals above the reservoir zone indicate late collapse. Late collapse displacement fracturing potentially allows oil and gas to escape to the surface, discrediting the prospect. Seek short interval isopach thins immediately above the reservoir target.

Reefs are usually encased in limestone formations. Specific density of reefs at 2.5 to 2.7 is slightly less than the surrounding limestone at 2.72. Therefore, a reef will isostatically attempt to "float" upward in limestone. Reefs forming in regressing (shrinking) continental basin seas are typically covered by gypsum with a specific gravity of 2.32. Consequently, that reef (2.5-2.7) will not float upward.

High-resolution seismic sends back a reflection each time a denser rock layer is encountered. Therefore, the top of the limestone (~2.72) at the base of the gypsum (~2.32) should produce a distinct seismic reflection. By comparison, the top of a porous (low-density) paleo-reef should be characterized by a less distinctive reflector than the surrounding limestone.

Clastic Exploration

During the Post-Flood Millennium, intermittent collapse below continental basin seas launched tsunami waves across adjacent lands. Lowland ecosystems were ripped out, pushed ahead, and subsequently dragged back in the tsunami wave backwash.

Tilted, lineament-bounded blocks formed low-elevation subaerial escarpments that diverted backwash sheetflow along its retreat back toward the continental basin sea. Backwash sheetflow eroded channels along the up-dip edge of raised blocks, forming lineament-controlled, zigzag patterns. Coarser grained sands were segregated by winnowing and deposited where those channels focused the most rapid backwash flow velocities. Dislodged organic materials tumbled along and were buried beside those discontinuous channel sands.

Examination of cores from Pennsylvanian Tyler Formation channel sand reservoir rock, from the Williston Basin of North Dakota, revealed an abundance of the mineral glauconite. Glauconite is a marine precipitate that is too fragile and too chemically unstable to be transported, deposited, and preserved. My partner, Dave Zwart, and I exchanged bewildered looks, as we immediately realized these Tyler channel sand cores must in some way be associated with marine waters. The presence of glauconite discredited the traditional view, that these channel sands had been deposited by fresh-water rivers.

Channel sands are located by mapping the up-thrown side of tilted lineament blocks that divert and concentrate retreating tsunami backwash along channels. Target thick Tyler reservoir isopach intervals on the up-dip side of the raised corner of each lineament block, where tsunami backwash was diverted and focused in narrow channels. Resulting higher flow velocities winnow out clay components, yielding porous reservoir quality sand. High-resolution seismic may detect a thickened reservoir zone due to actual thickening, plus slower seismic velocity through the porous reservoir sandstone.

Other Tyler reservoir sands parallel the Williston Basin rim in a position further toward the basin center. Because the previously-described channel sands were thought to represent on-shore river channels, these lower-permeability reservoirs were wrongly interpreted to be off-shore sand bars.

PETROLEUM EXPLORATION - RESERVOIR SAND DEPOSITION

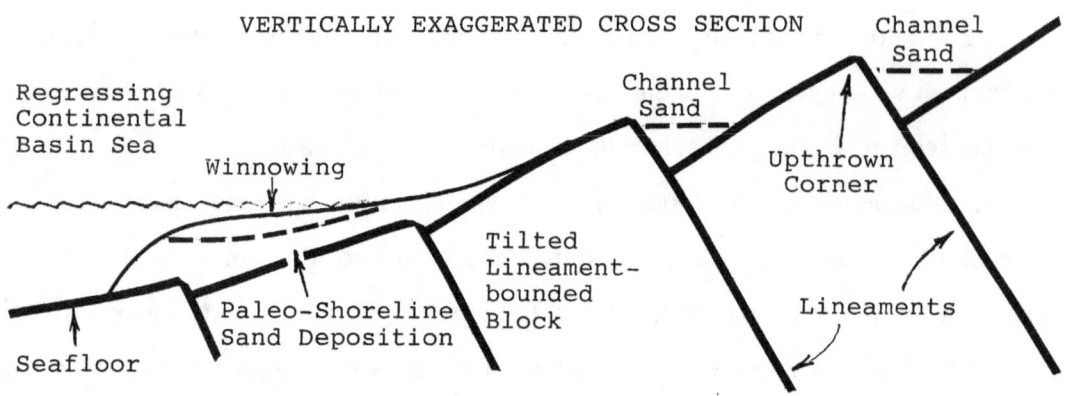

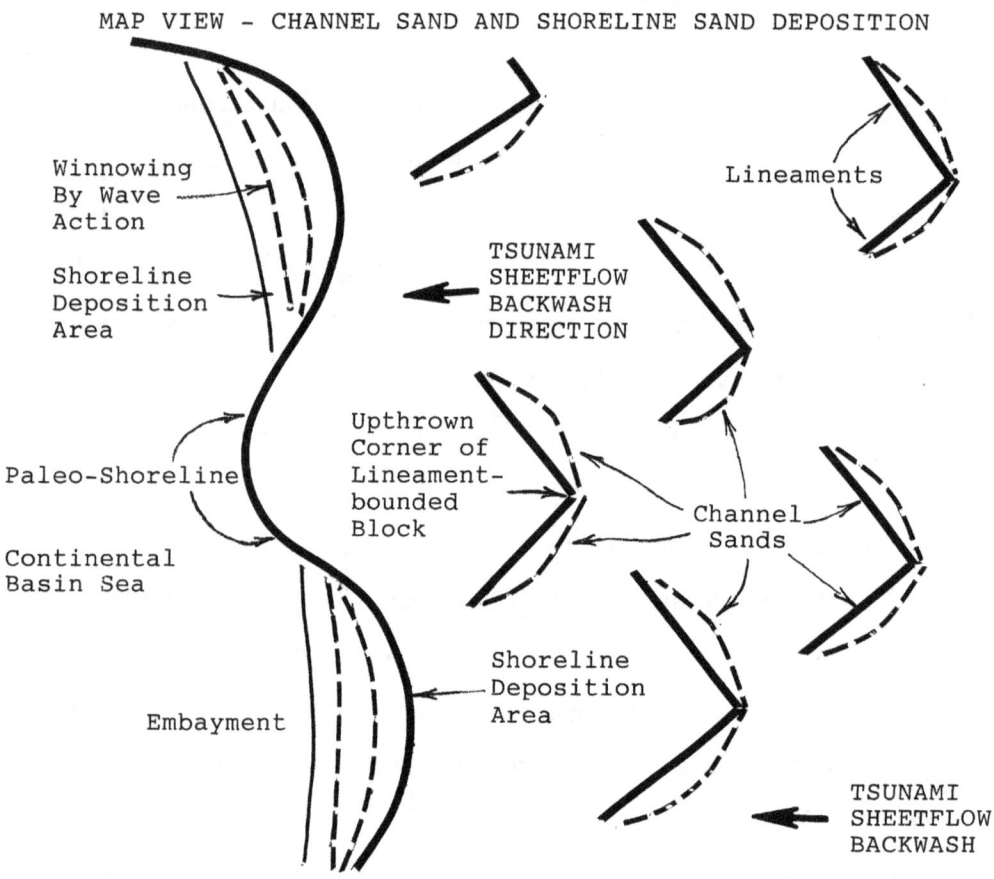

As the tsunami backwash entered the continental basin sea, flow velocities rapidly slowed, dumping a large volume of sediment mixed with organic material. The upper seaward portion of shoreline clastic deposits, were subsequently winnowed by wave action, removing clays and enhancing reservoir permeability. That reservoir rock was later covered by impermeable precipitates, evaporites, or muds before organic materials below could decompose and escape into the seawater above. Oil and gas were subsequently generated that migrated upward into the winnowed reservoir sandstone, displacing higher-density water, which had previously occupied porous intergranular cavaties.

Paleo-shoreline sands are mapped by laying out cross sections from basin center to rim. Reservoir isopach interval thickening toward the rim indicates potential for a tsunami backwash paleo-shoreline reservoir. Multiple similar cross sections of a specific reservoir zone should confirm this paleo-shoreline thickening to be oriented parallel to the basin rim in discontinuous segments. Segmentation occurred when lunar tides and/or river flow eroded narrow sections of reservoir quality sandstone.

Subtle embayments along the coastline of the ancient continental basin sea received the greatest volume of backwash sheetflow and organic materials. Identify the up-dip side of the gently sinuous reservoir line to conduct high-resolution seismic and specify prospect drill sites.

Multiple tsunami depositional events filled the continental basin. Therefore, each targeted channel or paleo-shoreline reservoir is typically located further toward the basin center and positioned higher in the stratigraphic column. Relative to existing oil and gas fields, exploration studies typically shift stratigraphically up and in toward the center of the basin, or shift stratigraphically down and outward toward the flank of the basin.

Each petroleum reservoir has a unique geomorphic history. Horizontal drilling techniques have enhanced petroleum recovery from low-permeability reservoirs where biogenic source rock and reservoir rock are the same (no petroleum migration).

COLLAPSE TECTONICS CRITIQUE

Collapse Tectonics advocates changing geomorphic processes through time. Scriptural information restricted interpretations to a Young Earth/Worldwide Flood geomorphic context. By restricting geologic time, many lower-probability alternative explanations were eliminated. Scientific accountability was maintained by prohibiting miracles-of-convenience from rescuing unworkable model parameters. Known geologic conditions and processes were proposed to compete head-to-head on a basis of scientific accountability with Naturalist counterpart models including Plate Tectonics.

Responsible model-building methodology requires the author to identify potential vulnerabilities of that model. Collapse Tectonics was constructed on two foundational postulates. Model parameters were developed from those postulates to establish sequential interdependency. Any successful attack would target Collapse Tectonics Model postulates.

Postulate 1 – advocates the large scale formation of porous crust. If the Antediluvian crust could be demonstrated to have been as fractured and permeable as the modern crust, extensive encapsulation of volatile liquids in vugs would have been too limited to generate enough porous crust to accommodate Collapse Tectonics Model parameters. **Postulate 2** – advocates vertical isostatic adjustment caused by changes in the specific density in Earth's crust. Isostasy is based on physics principles and is widely accepted as a viable mechanism causing tectonic movement. However, discrediting the continuing Post-Flood Millennium supply of *hypomagma* from continental roots would limit continental crust tilting and its corresponding effect on Phanerozoic sediment deposition.

Model replacement requires assembling a comprehensive tectonic model featuring superior driving mechanisms and better explanatory power. By comparison, the greatest advantage of the Collapse Tectonics Model over Plate Tectonics is its scientifically-viable driving mechanisms. The following addendums identify serious technical problems plaguing popular geomorphic interpretation concepts.

ADDENDUM 1: PLATE TECTONICS

Plate Tectonics has reigned as the favored geomorphic model for over 50 years. The earlier Geosyncline Model was limited to regional applications. The Expanding Earth Model lacked a reasonable mechanism to cause such large scale expansion. **Consequently, Plate Tectonics survived for five decades as the preferred geomorphic interpretation option.**

Mantle Convection Cells

The primary Plate Tectonics deficiency is its inadequate mantle convection cell driving mechanism. Changing temperature induces relatively minor density variations in individual minerals. By comparison, density variations between different minerals are much greater. Therefore, uniform mantle density is required to allow differential temperatures to drive mantle convection cell flow. However, varying density minerals relentlessly stratify in any flowing medium. Seismic signatures reveal mantle density stratification boundaries at 660 km, 1000 km, and 2000 km of depth (Hamilton, W.B., 2003, pp. 4, 5, 11). More subtle density stratification layers are anticipated in the asthenosphere (a.k.a. low-velocity zone) and below, wherever material conditions enable flow. Mineral density segregation discredits the concept of uniform mantle density.

According to seismic wave behavior, rock is present all the way down through the mantle to the outer core boundary. Convection cell flow would be prevented by solid mantle materials.

The dilemma for Plate Tectonics theorists is to reasonably explain how large scale mantle cell convection flow can exist in the face of contrary evidence. Plate Tectonics authors including Allan Cox were painfully aware of this proposed driving mechanism deficiency. Cox posed an extra credit question to his Stanford University students requesting a viable alternative to mantle convection cells (Bracken, 2012, pers. comm.). Mantle evidences supporting a scientifically-viable driving mechanism have eluded Plate Tectonics theorists since then.

Geologic evidences of compression, tension, upheaval, and subduction are simply assumed to have been generated by crustal displacement driven by mantle convection cell motion. However, geomorphic interpretations can also be explained by geologic processes unrelated to Plate Tectonics. Using geologic evidences to justify a model postulate is like *"the tail wagging the dog."* It first assumes what it intends to justify by secondary evidence (circular reasoning). **Plate Tectonics Theory is discredited by the failure of mantle convection cell motion as a reasonable driving mechanism.**

Earth Age

In geomorphic modeling, the shortest possible time frame explaining a geologic process provides the highest probability of a correct interpretation. In contrast, Plate Tectonics adopts a virtually unrestricted time frame, allowing the proliferation of lower-probability explanations. As a result, Plate Tectonics theorists commonly attribute contradictory forces to explain the same geologic evidence without resolution.

By comparison, Collapse Tectonics Model parameters are restricted to an orders-of-magnitude shorter time frame. That shortest possible time frame modeling technique maximizes the likelihood of correct interpretations, but in no way restricts Earth Age from being expanded thereafter to conform to other evidences.

ADDENDUM 2: RADIOMETRIC AGE DATING

Old Earth age dates are rooted in the assumed reliability of Radiometric Decay Dating Methods. Radioactive isotopes have an unstable nucleus that decays into stable elements by emitting energy and nuclear particles. The original unstable, radioactive form is known as a "parent isotope". The product elements (including the final stable form) are known as "daughter isotopes". Radioactive Decay Dating Methods compare the proportional quantity of radioactive parent isotope to the quantity of the final stable daughter element. The calculated time required for half of a given quantity of radiometric parent isotope to decay into its stable daughter product determines its "half-life". The half-life of each radioactive isotope is extrapolated back through time to a presumed starting point. A smaller proportion of parent isotope signifies an older extrapolated date. **Each Radioactive Decay Dating Method is subject to a unique set of variables capable of introducing error.**

CARBON-14 DATING

Carbon-14 Dating is used to date formerly living materials back through time for thousands of years. Cosmic rays strike the nuclei of nitrogen atoms in Earth's atmosphere.

Two beta particles (electrons) are absorbed by the nitrogen atom nucleus, changing two protons into two neutrons, thereby producing radioactive carbon-14. Under normal conditions, it takes 5,735 years for one-half of the original carbon-14 to decay back into nitrogen. Therefore, older plant and animal remains usually contain a lower percentage of carbon-14 than more recent organic material. If preserved organic matter contains one-half the modern carbon-14 ratio, the sample is dated at 5,735 years old. If the ratio decreases to one-forth of the modern ratio, the sample is dated at 11,470 years. One-eighth of the modern ratio is dated at 17,205 years, etc. Confidence is increased when the time of recorded measurement is long compared to the duration of extrapolated time. Due to a comparatively short half-life, the Carbon-14 Dating method provides a greater confidence level for extrapolated dates than its longer half-life isotope cousins.

Reasons for incorrect carbon-14 dates may include bacterial activity (Morris, 1994, p. 65), organic material types, water temperature, and chemicals within the water. Freshly-killed fur seals were dated at 1,300 years. Mummified seals dead approximately 30 years were dated at 4,600 years (Amer. Jour. of the U.S., 1971, p. 210). Living snails in artesian springs were dated at 27,000 years (Riggs, 1984, pp. 58-61). Different parts of the same baby mammoth adjacent to a stream dated 26,000 years old and 40,000 years old. Wood buried beside that baby mammoth dated at 10,000 years (Science, 1984, Vol. 224, p. 3). Conventional wisdom implies C-14/C-12 differences in the atomic nucleus should not cause selective leaching. However, the association between sample exposure to water and C-14 depletion resulting in excessively old age date extrapolations is uncanny. Arid climates provide the best C-14 dating archaeological correlations.

"Much of the groundwater found across the earth contains naturally-occurring ^14C in varying concentrations ... ^14C continues to flow through the world's aquifers even today, and anything removed from the subsurface saturated by groundwater would also contain traces of naturally-occurring ^14C. We would suggest that ^14C would be expected almost everywhere saturated materials would be subject to subsurface groundwater flow. Hence, finding traces of ^14C in coals of Mesozoic and Tertiary age would be expected based on our present state of knowledge of ^14C in the world's aquifers" (Froede and Akridge, 2013, p. 333). Therefore, trace amounts of C-14 provide an inappropriate measure of formation age where groundwater exposure has occurred.

The author of radiocarbon dating, Willard Frank Libby, determined that according to his measurements, more carbon-14 is being produced in the atmosphere than is decaying into carbon-12. Extrapolating backwards through time, Libby calculated carbon-14 would have been absent from Earth's atmosphere 16,000 years ago. Because Libby assumed Earth's atmosphere is billions of years old, he believed these measurements were a temporary local effect, and chose to ignore increasing atmospheric content of carbon-14 (Nutting, Dave, 1994, Seminar; Bailey, CO).

The production rate of carbon-14 exceeds the natural decay rate by as much as 25% (Lingenfelter, 1963, p. 51). Cook and Whitelaw conducted independent studies of the increasing content of carbon-14 in Earth's atmosphere based upon more extensive data than Libby's. Both came to the conclusion that carbon-14 would have been absent from Earth's atmosphere as recently as 8,000 years ago.

The Sun is a variable star. High sunspot/facula activity generates a greater production of carbon-14. In contrast, reduced sunspot activity during the 1600's and 1700's produced lower amounts of carbon-14. This effect shows up in tree ring data (Holroyd, E., 1999, pers. comm.). Atmospheric carbon-14 concentrations appear to directly correspond to cosmic ray bombardment from the sun and the decreasing intensity of Earth's magnetic field, which shields the atmosphere from those solar-sourced cosmic rays. Carbon-14 concentrations in living organisms are proportional to atmospheric carbon-14 concentrations. **Adjustments for less atmospheric carbon-14 in older samples commonly yield extrapolated dates very close to Biblical chronologies and documented historical accounts at archaeological sites in arid climates.**

Earth's Decreasing Magnetic Field

Increasing carbon-14 in the atmosphere appears to be affected by Earth's decreasing magnetic field. It is believed that a swirling motion of Earth's liquid iron core generates the magnetic field. However, the choreography of this motion and what energy sources drive it remain unexplained (Weisburd, 1985, p. 218). We do not know of any mechanism by which Earth's magnetic field might be "recharged".

Magnetic Field Satellite (Magsat) data indicates Earth's magnetic field is declining at a rate of 26 nanoteslas per year. If that rate of decline continues, magnetic field strength would decline to insignificance in 1,200 years (Anonymous, 1980a, p. 407). Testing of archaeological artifacts confirms Earth's long-term magnetic field decay. *"Dipole decay is evident in Roman ceramic artifacts, which contain iron particles that are magnetized to a greater extent than are those of modern products"* (Bloxham and Gubbins, 1989, p. 71).

Historical data implies Earth's magnetic field is decaying at a 1,400-year half-life. Extrapolating back through time, Earth's magnetic field would have been comparable to a magnetic star 100,000 years ago. Thomas Barnes has speculated that Earth's internal structure would have been disrupted by heat produced 20,000 years ago (Morris, 1994, p. 75). Also, life as we know it would not have been possible in the not-too-distant past. **Prehistoric life on Earth may have been limited to tens of thousands of years.** At its present rate of decay, Earth's magnetic field will for practical purposes cease to exist around 10,000 A.D. (Morris, 1994, p. 75).

LONG HALF-LIFE ISOTOPE DATING

Long half-life radiometric dating methods yield extrapolated dates ranging from hundreds of thousands to a few billion years in age. These methods target the time of rock solidification. (Melting resets the radiometric decay clock.) Igneous and metamorphic rocks that contain a sufficient concentration of long half-life radioactive isotopes can be dated. Sedimentary rocks rarely contain enough radioactive isotopes to date the samples. Also, clastic sedimentary rocks are composed of minerals that solidified at some unknown time prior to being eroded, transported, and deposited. Therefore, sedimentary formations are radiometrically dated using interbedded lava flows, volcanic ash beds, and intrusive dikes and sills.

During the mid-1900's, it was assumed that isotope half-life decay was controlled exclusively by random fluctuations in the atomic nucleus. (Processes outside of the atomic nucleus were presumed not to affect isotope decay.) However, laboratory evidence had already revealed radioactive decay rates are altered by extra-nuclear forces including pressure, temperature, electric and magnetic fields, and stress in monomolecular layers (Emery, 1972, p. 165). However, those outside-induced decay rate accelerants were ignored because those error factors were considered small enough to fall within the margin of error for radiometric calculations. Also, there was no reliable means of quantifying outside-induced decay rate accelerants in nature.

Binding energies within the isotope nucleus restrain the decay process. Higher binding energies per nucleon correspond to longer half-life isotopes. Disruptions must be of sufficient energy to destabilize the binding energy, thereby initiating alpha or beta decay.

The recognition that forces outside the isotope nucleus are capable of accelerating decay calls into question the extent to which random fluctuations in the isotope nucleus control decay rates. Random subatomic fluctuations potentially provide a reasonable cause of unstable short half-life isotope decay rates measured in milliseconds to days. However, random fluctuations in the nucleus may not be capable of destabilizing stronger binding energies of more stable isotopes. Isotopes with half-lives measured in millions to billions of years may be predominantly controlled by forces outside the isotope nucleus.

Isotope decay rates in nature may fluctuate as time passes in response to changing destabilizing forces exerted from outside the isotope nucleus. Each radiometric decay dating method loosely plots a trend with the majority of extrapolated dates corresponding to the same order as Geologic Column stratigraphy. The youngest extrapolated dates are typically recorded at the top of the Geologic Column, progressing to successively older dates deeper in the Geologic Column. Various decay-accelerating processes affect each isotope differently. Consequently, it is common for several different isotopes in the same rock sample or formation to yield a wide range of extrapolated decay dates.

The reliability of extrapolated radiometric decay dates resides in the assumption that laboratory testing is representative of isotope decay rates in nature. To accommodate that assumption, the observed widespread scatter of extrapolated radiometric decay dates has led to a proliferation of acknowledged open-system parent/daughter ratio variables. However, the widespread scatter of extrapolated radiometric decay dates is more simply explained by intermittent isotope-destabilizing decay-accelerating processes in nature. *"Isotopic Dating remains overloaded with numerous layers of assumptions, special pleadings, and selective manipulation of data ... Any unwanted result can be explained away, at will, on an after-the-fact basis ... isotopic dates which don't agree with the then-current* *ideas as to correct age are summarily rejected"* (Woodmprappe, 1999, pp. 8, 38, 95**.**). Let us consider a variety of extra-nuclear processes that potentially explain the difference between closed-system laboratory decay rates and open-system isotope decay rates in nature.

Neutrino-induced Fission

Neutrino-induced fission generates a very low level of background decay that accompanies other more rapid forms of decay in nature. Neutrinos are expelled as a byproduct of the nuclear fusion that powers the sun. Solids are composed of tiny, widely-spaced particles bound together by electromagnetic forces. Neutrinos are electrically-neutral and are therefore capable of traveling long distances through matter because they are not affected by charged particles such as protons and electrons. Therefore, an isotope nucleus in Earth's center is almost as likely to be struck by a neutrino as an isotope nucleus on Earth's surface facing the Sun.

A neutrino is a subatomic fermion particle with a half-interger spin. Flavor oscillations imply that neutrinos have an extremely tiny mass. *"Most neutrinos passing through the Earth emanate from the Sun. About 65 billion solar neutrinos per second pass through every square centimenter perpendicular to the direction of the Sun in the region of the Earth …Very much like neutrons do in nuclear reactors, neutrinos can induce fission reactions within heavy nuclei"* (Wikipedia, "Neutrino," p. 1).

Neutrino bombardment fluctuates according to temporary variations in solar output. Neutrino-induced fission reactions *"populate the daughter nucleus at excitation energies where the shell effects are significantly washed out, effectively reducing the fission barrier"* (Kolbe, et al., 2008, p. 1). *"In general the interaction probability increases with the number of neutrons and protons within a nucleus"* (Wikipedia, "Neutrino," p. 3). When neutrinos do strike an atomic nucleus, *"… the energy reduction is significant, amounting to about 4 MeV on the average"* (Kolbe, E., et al., 2004, pp. 2-3). That average energy reduction is sufficient to initiate beta decay and some alpha decay (Boudreaux, 2012, pers. comm.).

Data from the Brookhaven National Laboratory on Long Island and the Federal Physical and Technical Institute in Germany revealed silicon-32 and radium-226 measure a slightly faster decay rate in winter than in summer (Stober, Dan, 2010, pp. 1-2). Earth's elliptical orbit results in a slightly closer separation from the sun in winter, causing a slightly more concentrated dose of neutrinos. There is also a discernible and persistent chlorine-36 decay rate fluctuation of less than 1% between the aphelion and perihelion of Earth's orbit (see A.A.B. & J.P., 2012, p. 1). Also, two separate experiments at two different labs have confirmed changes in the decay rate of the chlorine-36 isotope corresponding to solar flares.

Widespread variations in radiometric dating extrapolations beg for the identification of specific decay-inducing cause-and-effect mechanisms that uniquely affect different isotopes. (The primary cause of long half-life isotope decay under laboratory conditions may not be random fluctuations in the atomic nucleus as previously believed.) Neutrino-induced fission is a viable candidate for initiating high-density nucleus isotope background decay as measured in laboratory tests. However, laboratory conditions can only duplicate neutrino-induced background decay in nature. **In contrast, laboratory half-life decay rates are potentially orders-of-magnitude slower than open-system half-life decay rates in nature.**

Plasma State Accelerated Decay

Plasma or depleted electron conditions not mimicked by standard laboratory half-life tests may be lowering the isotope binding energy threshold by orders-of-magnitude one atom at a time. Consider what might routinely occur in nature each time electron flow passes across an isotope. Some or all of the electrons surrounding the isotope nucleus are temporarily absent, resulting in a brief pulse during which the nucleus is more vulnerable to beta decay. However, it is the outer electrons that are more likely to be temporarily absent. *"Beta decay rates would be governed more by the inner electrons, as those are closest to the nucleus"* (Wile, 2013, pers. comm.). Therefore, electron flow would affect each isotope differently according to relative binding energies of surrounding electrons and intensity of current flow. That in turn would potentially affect beta decay rates. *"I would think that if local plasma affects are the cause of accelerated decay, you would see a lot more scatter in the results of radiometric dates by location. Some rocks, for example, should conduct electricity better than others, making some rocks more susceptible to the acceleration. This would be a way to test the idea"* (Wile, 2013, pers. comm.)

"The foregoing discussion assumes that electrons surround the nucleus, which of course is nearly always the case. For over 50 years, however, some theoreticians had suggested that negatron (beta) decay could be altered in the case of a nucleus bereft of its electrons (as occurs in a plasma state). Perhaps the B- particle attempting to leave a bare nucleus would have to overcome a much lower threshold of kinetic energy than if the electrons were present. The fleeting B- particle could take refuge in a vacant electron orbital around the nucleus instead of attempting to escape all the way into the continuum. This process is called bound-state B- decay ... Subsequently, theoretical annalyses (see Takahashi, et al., 1987, pp. 1522-1527) suggested that a significant perturbation of radioactive decay rates could occur in the nuclides of 25 different elements as a consequence of (bound-state) *B decay* (Woodmorappe, John, *creation.com*, "Billion-fold acceleration of radioactivity demonstrated in laboratory," pp. 1-3.).

High-density isotopes such as U-238, U-232, Th-240, etc. include both alpha and beta decay along their multi-step decay chains. How might these high-density isotopes be disporoportionally accelerated by intermittent electrical current-induced decay pulses that are only strong enough to initiate lower energy beta decay? *"It can be speculated that the collective energies of all these beta decay events might be sufficient to form a very localized, exceedingly highly energetic plasma (even for a very short period of time), sufficient to accelerate even the alpha decay"* (Boudreaux, 2013, pers. comm.). Any accerleration of longer half-life alpha decay stages would greatly accelerate the progression through the decay chain.

In contrast, C-14 samples typically consist of comparatively low density organic materials that behave as insulators, retarding electrical current flow. Consequently, electron flow would have had a negligible effect on inducing C-14 beta decay. **The orders-of-magnitude difference between C-14 and long half-life isotope extrapolations in associated samples may be caused by intermittent current-induced plasma-state decay selectively accelerating long half life isotopes.**

Piezonuclear Fission

The controversial concept of piezonuclear fission is in its infancy. If true, fault displacement, earthquakes, or overburden collapse events may temporarily generate shock compression of sufficient energy to induce fission decay. Three different kinds of compression tests were conducted on brittle rocks: *"(1) under nonotonic displacement control, (2) under cyclic loading, and (3) by ultrasonic vibration ... For specimens of larger dimensions, characterized by a brittle behavior, neutron emissions, detected by He^3, were found to be of about one order of magnitude higher than the ordinary natural background level at the time of the catastrophic failure"* (Carpinteri, et al., 2011, pp. 3-4). In the experiments on liquid solutions, aluminum atoms appeared at the end in a final quantity as large as about seven times the small initial quantity (Cardone and Mignani, 2007; Cardone, et al., 2009). Therefore, piezonuclear fission events may periodically accelerate radioactive decay rates.

Intermittent compression events spanning geologic time could periodically induce piezonuclear fission. **Older strata would potentially be exposed to more piezonuclear events, causing more extensive parent to daughter isotope fission.** The characteristic relationship of old to young age from bottom to top of the Geologic Column would be enhanced. However, the Geologic Time Scale as extrapolated by assuming laboratory rates in nature would potentially be orders-of-magnitude too old.

Critical Mass

Critical mass conditions accelerate fission decay exponentially by chain reaction. Energy released by isotope fission decay can shorten half-lives of adjacent isotopes. *"There is apparently virtually universal agreement that certain deposits found at the Oklo uranium mine in Gabon, Africa are the remains of "natural nuclear reactors ... Atoms of U-235 are naturally unstable and their nuclei undergo spontaneous fission at a very slow rate; however, if a U-235 nucleus absorbs a free neutron from its environment it usually fissions (breaks apart) virtually instantly ... Usually two or three excess neutrons are produced from the fission of each U-235 nucleus ... These neutrons are then available to trigger the splitting of other U-235 nuclei"* (Matthews, 2003, pp. 209-210). High concentrations of natural uranium apparently achieved critical mass, thereby initiating rapid chain reaction decay at Oklo.

Chain reaction decay also potentially functions on a microscopic scale. Consider a single biotite grain sample containing millions of atoms of parent isotope uranium-235. Isotope decay within the biotite grain instigates supplemental chain reaction decay in adjacent U-235 atoms, increasing the proportion of daughter lead. **Half-life decay rates in nature may be accelerated by isotope proximity from one atom to the next where intragranular isotope concentrations are extremely high.**

Recent Lava Flows Yielding Old Earth Ages

Mt. Ngauruhoe, New Zealand andesite lava flows deposited in 1945, 1959, and 1975 were extensively sampled, dated, and evaluated utilizing modern isochron methods. Seemingly valid isochrons yielded the following results:

1.) 5 point Rb/Sr extrapolated age of 133 +/- 87 million years.

2.) 5 point Sm/Nd extrapolated age of 197 +/- 160 million years.

3.) 7 point Pb 207/Pb 206 extrapolated age of 3908 +/- 390 million years.

(see Snelling, 2003, pp. 292, 299, 300.)

Carbon-14 age dates are frequently orders-of-magnitude younger than associated potassium/argon dates. In Crinum, Central Queensland, Australia, partially-fossilized wood was found imbedded in basalt. The wood had been partially burned by contact with the hot basaltic lava. The wood was carbon-14 dated at 44,800 +/- 950 years. The basalt was potassium/argon dated at 36.7 million +/- 1.2 million years (Snelling, 1997, pp. 24-27; Snelling, 2000, pp. 99-122).

Volcanics at Mt. Rangitoto, New Zealand, destroyed trees that were radiocarbon dated at less than 300 years. Radiometric potassium/argon dating estimated eruption age at 485,000 years ago (McDougall, 1969, pp. 1485-1520). Potassium/argon dating yielded apparently erroneous extrapolated dates orders-of-magnitude too old.

It is reasonable to expect that some daughter is present in most magma prior to solidification. If the magma is thoroughly mixed by rapid flow and solidifies quickly (before parent/daughter segregation occurs), the original daughter content of the magma cannot be identified. Unidentified daughter in lava is usually cited as the reason why modern lava flows consistently date orders-of-magnitude too old. However, ancient interbedded lava flows are used to date sedimentary formations throughout the Rock Record. **If you cannot trust radiometric decay dating methods to accurately date samples of known age, why would you trust that same methodology to date samples of unknown age?**

Parent/Daughter Ratio Alteration

A closed radiogenic system is initially assumed, meaning there has been no selective post-deposition supplementation or depletion of either the parent isotope or stable daughter element between the time of rock solidification and testing. *"Most, if not all, isotopic systems of Paleozoic age have been open, to one degree or another, for some period of time since their formation* (Kunk and Sutter, 1984, p. 13). In nature, the post-solidification variables in play are numerous and unquantifiable. Magma chamber zonation, isotope fractioning, magma mixing, extensive-term solidification, inherited isochrones, metasomatism, systems open to parent or daughter migration, hydrothermal alterations, etc. are routinely cited for the purpose of discarding data that fails to conform to desired Geologic Column age dates. In contrast, most extrapolated dates within the expected age range are routinely granted closed radiogenic system status. However, scientific accountability is automatically forfeited because that self-fulfilling affirmation first assumes what it is attempting to prove (circular reasoning).

Each radiometric method exhibits unique open-system vulnerabilities altering parent/daughter ratios. Radiometric decay dates may be extrapolated orders-of-magnitude too old if the parent isotope is depleted or the radiogeneic daughter is supplemented by metasomatic or hydrothermal conditions. Woodmorappe (1999, pp. 28, 33) lists the following examples and more. *"The U-Pb and Rb-Sr systems are known to be highly susceptible to resetting by hydrothermal, diagenetic and metamorphic processes"* (Toulkeridis et al., 1998, p. 138). *"Unfortunately, the parent radionuclide, 87-Rb is a volatile and mobile alkali element, characteristics which lead to open-system behavior and anomalous ages"* (Shirley, 1991, p. 11). *"Unfortunately, the U-Pb and Th-Pb systems rarely stay closed in silicate rocks, due to the mobility of Pb, Th, and especially U, under conditions of low-grade metamorphism and superficial weathering"* (Dickin, 1997, p. 105). *"There are many examples of intrusive granitoids containing old inherited radiogenic Pb in zircons which show an incomplete resetting of the U-Pb systems during melting events"* (Hong, et al., 1990, p. 133). *"A major complication limiting the* (U-Pb geochronology) *technique is the presence of xenocrystic zircon either incorporated from wall rocks during magma emplacement or entrained in magma as a residual unmelted phase from its source region"* (Roddick and Bevier, 1995, p. 307).

Do radiogeneic systems ever remain continuously closed in nature? The greatest potential for deposition of a pure radioactive parent isotope is produced by segregated leaching and exclusive parent isotope deposition at a different location. For example, acidic groundwater or metamorphic fluids selectively leach parent uranium from crystalline bedrock while leaving daughter lead in place. Uranium is then transported by groundwater to a secondary location. Uranium is attracted to and often replaces organic materials which contain no measurable lead. Therefore, uranium enrichment in bones can occur during the fossilization process.

Fossilization occurs when minerals from organic remains, including bone, are replaced by minerals from an outside source. Fossilization usually occurs soon after burial, thereby preventing bone decomposition. Dinosaur Ridge near Morrison, Colorado, U.S.A. contains uranium-rich fossilized dinosaur bones. Geiger counters are used to separate radioactive fossil bone fragments from surrounding non-radioactive sedimentary rock (Holroyd, E., 2000, pers. comm.). According to the Geologic Time Scale, these Jurassic fossils were presumably deposited and fossilized approximately 150 million years ago. **Radiometric Decay Dating tests on these dinosaur bones would identify the time of uranium infusion during the fossilization process.** Because fossilization probably occurred soon after burial, a radiometric age date of thousands of years would support a Biblical Young Earth Time Scale. A radiometric age date around 150 million years would support the Geologic Column Time Scale.

Isochrons

It is reasonable to assume that at least some *non-radiogenic daughter* from the original magma was present in most rocks when they solidified. Isochrons are a mathematical technique designed to correct for otherwise unrecognizable *non-radiogenic daughter* present in the magma source. A reliable isochron requires an unusual set of conditions. The original magma must contain a uniform distribution of parent, daughter, and sister daughter. Uniform distribution is facilitated by a thorough mixing of the original magma via high heat and rapid flow. That mixing must be coupled with the avoidance of external material contributions to the magma. Unfortunately, high heat and rapid flow are commonly accompanied by material contributions from external (non-magma) sources.

Chemical fractionation can also result in a single linear sloping line on an isochron graph. In a batholith, a lower-density daughter should segregate, migrate, and concentrate higher in a magma chamber than a higher-density parent that solidifies into crystals and sinks. Different melting points for the parent and daughter also produce chemical fractionation during prolonged solidification and during intermittent post-solidification heating events. Chemical fractionation processes produce abundant mixing lines that mimick a legitimate isochron (characterized by a single linear sloping line on the graph). However, mixing line isochrons are unrelated to original magma concentrations and therefore are misleading in determining the age of the rock.

Parents and daughters are also affected by different solubilities and by liquid/gas alteration at changing pressures and temperatures. Parents and daughters have differing tendencies to bind with other minerals or be trapped within crystal imperfections. Chemical fractionation processes within a magma chamber and during lava flows are too numerous and too unpredictable to quantify.

In conclusion, isochron calculations of daughter concentrations in the original magma must first assume the chemical fractionation processes above have not been in play. That idealized condition would occur very rarely if ever in nature. Consequently, a linear series of points on an isochron graph does not assure a valid measure of non-radiogenic daughter in the original magma. Unwanted isochron dates are routinely discarded and attributed to non-chronometric isochron plots, inherited isochron, apparent isochron, mantle isochron, pseudoisochron, secondary isochron, source isochron, erupted isochron, mixing line, and mixing isochron problems. **The selective acceptance or rejection of evidence based on desired results has evolved to make isochron methods non-falsifiable, thereby rendering its conclusions scientifically unverifiable.**

Naturalist Earth Age Estimates

The link between the Geologic Column and astronomical theory determines which discordant radiometric dates are assumed to be correct. In compliance with the proposed Big Bang event about 13.7 billion years ago, Naturalists propose hydrogen and helium gravitationally contracted, forming the Sun. Heavier elements are generated in the Sun. To obtain the heavy elements we observe on Earth, it is thought that our Sun is a third-generation star, following two previous novas and subsequent gravitational contractions, each forming a new Sun. The surrounding planets and meteorites are proposed to have formed from excess star fragments when our present Sun formed around five billion years ago (DeYoung, 1995, pp. 86-87.) Therefore, it is assumed that meteorites (from star fragments) are the same age as Earth.

Earth age estimates are based on variable radiometric decay dates of meteorites. Uranium/Lead and Thorium/Lead ratios in meteorites are extremely low. Consequently, radiometric methods yield extremely old extrapolated dates for meteorites. In fact, radiometric dates assuming pure radioactive parent elements are usually too old to allow the Sun to be a third generation star. Therefore, it is conveniently assumed that meteorites contained a parent/daughter mixture at time Earth formed.

The Canyon Diablo Troilite meteorite's parent/daughter element ratio was selected as a "correction factor" to revise radiometric dates corresponding to astrophysics theory. The Canyon Diablo Troilite meteorite is assumed to contain a representative ratio of radiogenic (radioactively-decayed) to non-radiogenic (undecayed) lead. Also, the Canyon Diablo Troilite meteorite is assumed to contain the same parent/daughter element ratio as a stony meteorite named "Allende". How do Naturalists know that? They don't! When the meteorite solidifies, any isotope/daughter ratio is possible. The meteorite Allende contains many radioactive elements in sufficient volume to be radiometrically dated. Rb/Sr techniques yielded ages from 0.70 to 4.49 billion years for inclusions. Surrounding rock matrix ages by Rb/Sr were reported at 4.60 to 4.84 billion years. Other radiometric techniques included Sr-87/Sr-86 at 4.48 billion years, Pb-207/Pb-206 at 4.50 billion years, U-235/Pb-207 at 5.57 billion years, U-238/Pb-206 at 8.82 billion years, and Th-232/Pb-208 at 10.4 billion years (Morris, 1994 p. 61; Gale, et al., 1972, p. 57).

The preceeding topics have focused on explaining why isotope decay rates measured under laboratory conditions do not duplicate isotope decay rates subjected to alteration by various subsurface conditions. The same is true of meteors in space. For example, greatly increased cosmic ray exposure acting in concert with the absence of a significant gravitational field might accelerate isotope decay by several orders-of-magnitude.

It is unreasonable to use extremely speculative radiometric extrapolations derived from meteorites as a basis for determining the age of the Earth. By extension, Geologic Column age dates are adjusted accordingly to supposedly provide sufficient time for Biologic Evolution to occur.

ADDENDUM 3: THE GEOLOGIC COLUMN

"The geologic column is not as much read from the rock as it is read into the rock."
(Woodmorappe, 1980, p. 212)

The Geologic Column construct is the prehistoric template of old-Earth Naturalism, and the handmaiden of Evolution Theory. Fossil ecosystems preserved in the Rock Record are treated as "snapshots" depicting stages of Biological Evolution during separate periods of geologic time. Fossil successions and lithologic sequences exhibit a similar vertical order in continental basins throughout the world. Biological Evolution is employed as the "guiding light" and *circular reasoning* is utilized as a self-affirming strategy to categorize fossil evidences within separate time capsules.

Sediment deposition occurs intermittently rather than continuously. Therefore, no sedimentary basin contains a complete rock column, representing Earth's entire depositional history. Distinctive sedimentary rock layers called "formations" are described in sequence from oldest at the bottom to youngest at the top to derive the Geologic Column.

Paleontological evidences are correlated across gaps between similar separate sedimentary formations and across oceans to assemble a presumed worldwide sequence of geologic time.

The *Principle of Superposition* states that younger rocks are deposited on top of older rocks. Therefore, vertically stacked sedimentary layers reveal a progressive sequence of prehistoric depositional events from any point below Earth's surface directly upward. The Geologic Column assigns each sedimentary formation in a continental basin to a unique depositional time-frame. Visualize a deck of playing cards stacked in separate piles. The deck represents the entire Geologic Column. Separate card piles represent different continental basins. Individual cards represent sedimentary formations. The Geologic Column is a theoretical attempt to integrate the card piles together into one complete deck or rock column that explains prehistoric Earth by assigning specific time slots to each card.

Although the vertical sequence of sedimentary formations exhibiting similar fossils and lithology is similar from one continental basin to the next, these separate formations correspond directly to characteristic paleoenvironment settings rather than unique geologic time periods. The indirect correlation between paleoecosystems and unique periods of geologic time must first assume Biologic Evolution to derive the Geologic Column.

The *Principle of Continuity* unreasonably assumes similar climates and depositional environments were taking place at the same time worldwide throughout geologic history. Furthermore, sedimentary formations from one continental basin are assumed to be time-equivalent to similar formations that occupy similar vertically-stacked positions in the next basin and on the next continent. Distant unconnected lithologies exhibiting similar fossil ecosystem biota and lithology are presumed to be time-equivalent if they occupy a similar stratigraphic position in the Geologic Column.

The *Principle of Continuity* ignores the extensive variety of unique ecosystems and sedimentary deposition conditions observed on Earth's surface today. Several different modern sedimentary formations are commonly observed to be actively deposited at any specific time across adjacent areas as continental water bodies shift. Flume experiments show that sediments transported by flowing water are deposited laterally through time (Berthault, 1997 pp. 65-70). Correspondingly, the overlapping "shingled" arrangement of sedimentary formations is observed in both modern and ancient deposits. Modern and ancient sedimentary formations were usually deposited laterally, as shallow seas transgressed and regressed back and forth across the land. Similar vertical fossil and lithology successions only indicate that a similar sequence of environmental and deposition changes took place. The deposition sequence of similar fossils and lithology occurred at different rates during different times in separate continental basins worldwide. Therefore, the Geologic Column nomenclature correlating directly to bracketed capsules of geologic time is rejected.

The Geologic Column alternative advocates that the proliferation and demise of these unique paleoecosystems was fostered by a typical worldwide progression of changing geologic and climactic conditions. Many paleoecosystems were present on Earth's surface at any specific time in Earth history. Lateral migration and deposition overlapping of these extinct paleoecosystems generated the stratigraphic sequence as observed. Earth's surviving lifeforms are the product of natural selection in changing ecosystems - not Biologic Evolution.

Worldwide Unconformities

Unconformities are erosional surfaces that truncate geologic formations below. Major unconformities supposedly represent extensive gaps in the Rock Record. Unconformities separating sedimentary formations are assigned as much time as necessary to expand the Geologic Time Scale to accommodate the *"deep time"* assumed to accompany Biologic Evolution Theory.

Erosional unconformity surfaces also shift laterally through time. Nevertheless, major unconformities covering large portions of sedimentary basins are correlated to other major unconformities at similar stratigraphic positions in other continental basins worldwide. Major unconformities are used to sort formations from different sedimentary basins into supposedly time-equivalent "periods". The misleading Geologic Column assumption is that these "major erosion events were all taking place at the same time in different parts of the world." However, erosion in one area is followed shortly thereafter by deposition of similar volume of material in an adjacent area. Therefore, it is unreasonable to expect that continental basins experienced lengthy periods of worldwide erosion and non-deposition at the same time.

Phanerozoic sedimentary formation boundaries can rarely be tracked very far without discovering angular unconformable surfaces. These unconformable surfaces frequently divide the formation into overlapping "shingles". In most cases, the formation was deposited by a progressive series of similar deposition events through time. **In conclusion, Phanerozoic unconformities usually represent time-transgressive, laterally-shifting erosional surfaces unrelated to counterpart unconformities in other continental basins.**

Paraconformities

We commonly observe sedimentary layers in contact that were supposedly deposited millions of years apart according to the Geologic Time Scale. Therefore, an unconformity is designated to separate the two layers. We would expect a surface exposed for millions of years to be eroded and rough with a soil zone and/or a weathered zone below the unconformity surface. Frequently, however, this supposed unconformity contact between layers is parallel with no sign of a soil layer or weathering of the underlying formation. **"Paraconformity contacts" appear to be simple bedding planes, where one layer was deposited almost immediately after the other.**

Uniformitarianism

Uniformitarianism is based on the foundational assumption that, *"the present is the key to the past."* Slow and gradual deposition processes observed in modern times are used as a measuring rod to extrapolate back through time. However, during the 1950's and 1960's authors including Velikovsky (1955) cited numerous catastrophic evidences that could not be reasonably explained by modern geologic processes, thereby discrediting strict uniformitarian interpretation.

The modern Creation Science movement was launched when Whitcomb and Morris (1961) accelerated the demise of uniformitarian thinking by advocating prolific Phanerozoic fossil evidences to have been catastrophically-deposited during the Biblical Worldwide Flood. The strength of that Creationist argument was in contrasting a slow and gradual uniformitarian background with abundant fossil evidences in most Phanerozoic strata. Since then, most Creation Science time and treasure has been dedicated to supporting, defending, and promoting the concept of most Phanerozoic strata correlating to Worldwide Flood deposits.

During the late 1900's abundant catastrophic evidences deep in the Rock Record established past cataclysms unlike geologic processes observed in modern times. Strict uniformitarianism morphed into Neo-catastrophism. *"In other words, the history of any one part of earth, like a soldier, consists of long periods of boredom and short periods of terror"* (Ager, 1993, p. 141). (The concept being that modern geologic processes correspond to *"long periods of boredom".*) Naturalists now acknowledge intermittent catastrophic depositional events in the past (except for a Worldwide Flood). The 1960's Creationist argument that catastrophic evidences correlate directly to Worldwide Flood deposits was no longer relevant. **Geologic context and scientific-accountability replaced uniformitarian assumptions as controlling factors in the Naturalist/Creationist debate.**

Precambrian/Tertiary Interfingering

Interfingering of distinctive sediments occurs when sediments transported from different source areas collide. While climbing a private driveway at 30644 Ouray Road near Pine, Colorado, my eyes locked onto a road cut revelation. The low-grade metamorphosed "Precambrian Pikes Peak Granite Grus" contained an interfingered tongue of distinctive "Tertiary Gravels". My initial thought was the Granite Grus was eroded from the Pikes Peak Granite and interfingered with the Tertiary Gravels at a much later date before low-grade metamorphism ensued. However, that notion was discredited, as subsequent examination of hundreds of excavated trenches revealed well consolidated Pikes Peak Granite grading with no apparent erosional contact upward, downward, and laterally into Pikes Peak Granite Grus. The degree of metamorphism varies dramatically across inches and appears to be part of one continuous rapid-cooling event. If one billion year old Pikes Peak Granite is interfingered with one million year old Tertiary Gravels, their time of deposition overlapped. **The Geologic Column Time Scale is irrelevant if 999 million years are missing from the Rock Record.**

BIO-STRATIGRAPHY

The assignment of prehistoric time to the Rock Record is called the "Geologic Time Scale", which is presented like a stratigraphic storybook, whose pages represent prehistoric dates. However, rocks do not come labeled with dates. Therefore, stratigraphic observations of the Rock Record reveal deposition sequencing – not time. In contrast, the Geologic Time Scale reverses that process by pigeon-holing geologic formations in time-denominated slots. **Consequently, Geologic Column terminology can only be applied to stratigraphic sequence - not assumptions regarding age.**

Transitional Fossils

If Biologic Evolution were true, the fossil record should provide a multitude of plant and animal fossils that transition continuously from one phylum, class, order, and family into the next. However, each widely touted example of a transitional fossil has been routinely disproved and quietly discarded. Instead, we observe distinct, fully formed kinds of plants and animals throughout the fossil record.

Dr. Colin Patterson, an Evolutionist and Senior Paleontologist at the British Museum of Natural History in London, studied transitional fossils for over twenty years. The following is an excerpt from a letter Dr. Patterson wrote to a fellow scientist on April 10, 1979. *"I fully agree with your comments on the lack of direct illustration of evolutionary transitions in my book. If I knew of any, fossil or living, I certainly would have included them. You suggest that an artist should be used to visualize such transformations, but where would he get the information from? I could not honestly provide it, and if I were to leave it to artistic license, would that not mislead the reader?"* (Sunderland, 1988, p. 89)

Empirical evidence confirming transitional fossils at the family level (a.k.a. kind) and above is absent. Some evolutionists (including S.J. Gould and N. Eldridge) acknowledge this lack of transitional fossils and propose a modification called *"Punctuated Equilibrium"*. This position proposes Biologic Evolution took place in brief intermittent pulses separated by long periods of negligible evolutionary advance. This tactic conveniently circumvents the need transitional fossils as supporting evidence for Biologic Evolution. However, the introduction of new species through massive, short-term beneficial changes is genetic fantasy.

Ken Ham states, *"One evolution model is too slow to see and the other is too fast to see."* One expert on fish expressed his frustration at the disturbing absence of evidence supporting Biologic Evolution. *"I have often thought how little I would like to have to prove organic evolution in a court of law"* (White, E. 1966, p. 8). **The absence of reliable transitional fossil evidence discredits the assumption that "Molecules-to-Man Evolution" can be applied as a Geologic Column construct.**

Cambrian Explosion

Cambrian fossils include over 50 animal phyla. Today, only 38 phyla of those animals are still living (Chien, 1997, p. 2). Every phylum of animal life appears abruptly in the Rock Record in Cambrian rocks. Each organism is *"fully formed and fully functional without less adapted ancestors in lower levels ... these diverse forms continue up the (geologic) column (i.e., throughout time) with much the same appearance possessed at the start ... The fossils at the bottom (i.e., long ago) are equally as complex as any animal today"* (Morris, 2003, p. d). **The sudden proliferation of Cambrian fossil evidences undermines the concept that simple life-forms evolved through time by "accidents of nature" into progressively more complex life-forms.**

Iterative Evolution

Naturalists believe "index fossils" evolved, proliferated, and became extinct during a brief duration of geologic history. Each time-bracketed index fossil implies the host rock was deposited during a brief age span specified by the Geologic Time Scale.

Naturalists are embarrassed when index fossils are found in host rocks outside the presumed time-bracket assigned to that species. The logical response would be to expand the time bracket for that index fossil. However, the value of an index fossil resides in a supposedly short time-bracket during its presumed evolution history. The number of suitable index fossils is small. Expanding the time bracket for an index fossil calls into question all the host rocks previously dated by that index fossil. That in turn undermines many presumptions upon which evolution family trees are based. Naturalists don't want that problem. So they propose a process called *"Iterative Evolution"*.

Iterative Evolution claims that the same index fossil form supposedly evolved repeatedly during its evolution history. Tiny foraminifera shells of various species are considered to be reliable index fossils. However, identical foraminifera shells are often assigned to a different genus or species depending upon the presumed age of their host rocks (Tosk, 1988, pp. 8-18). The index fossil conveniently dates the host-rock, and the host-rock verifies the index fossil species by circular reasoning. If Naturalists did not need the species for an index fossil, paleontologists would presume that foraminifera survived instead of becoming extinct and re-evolving a second time.

Adding to the speculation, foraminifera sometimes display different chamber characteristics under different ecological conditions. For example, different water temperatures result in different shell shapes for the same species. Some supposedly extinct forms may be varieties of foraminifera living under different ecological conditions (Oard, 1990, pp. 185-186). **Iterative Evolution provides unrestricted license to force-fit contrary evidence into the desired Geologic Time Scale age.**

Overthrust Belts

It is amusing to find ad hoc explanations proposing that older rocks have been shoved laterally for tens of kilometers across the top of supposedly younger rocks along what are termed "overthrust belts". Friction should pulverize a thick breccia layer (fault gouge) along the sliding surface of contact. However, there is rarely any breccia along the overthrust contact surface to imply displacement ever occurred. Overthrust contact surfaces are commonly described as *"knifesharp"* (Gretener, 1972, p. 590), *"fully conformable"* (Rezvoy, 1971, p. 1736), or as a *"mere bedding plane"* (Gretener, 1977, p. 111). Overthrust contact surfaces appear to have been laid one on top of the other in rapid succession. This exposes the inappropriate explanations that arise when false presumptions regarding the time-bracketing of microfossils are assumed as fact.

Interfingering of rock layers indicates simultaneous deposition of different rock-types. Lenses or "tongues" of one rock-type frequently protrude into another rock-type across supposed overthrust contact surfaces. This implies deposition was continuous from the rock layer below to the rock layer above, discrediting overthrust displacement along a planar slide surface.

Folds in bedding planes are usually continuous above and below the overthrust contact surface. That implies folding occurred before the sediments had time to harden (consolidate) into rock. However, unconsolidated material would not be brittle enough to be shoved over the top of the unconsolidated layer below.

"Gravity-gliding" is the mechanism used to explain the movement of one block of crust sliding across the top of another. An incredible fluid pore pressure must be maintained along the sliding contact surface to accomodate *"gravity-gliding.* "*We suspect that over the areas of large thrust sheets such as those in the Appalachians or the western Cordillera, effective permeabilities would have been too large to allow gravity gliding, even with shale or evaporite cap rocks"* (Guth, et al., 1982, p. 611). In addition, overthrust movement should generate massive fracturing in the overriding block, generating pressure release conduits. *"Consequently, for the conditions assumed, the pushing of a thrust block, whose length is in the order of 30 km or more, along a horizontal surface appears to be a mechanical impossibility"* (Hubbert and Rubey, 1959, pp. 126-127). Tectonic evidence for overthrust ranges is disturbingly weak. **The overthrust mechanism is commonly imposed for the sole purpose of reconciling out-of-place index fossils.**

Reworked Index Fossils

Index fossils from supposedly different periods of time in the Geologic Column are frequently mixed together. The out-of-place fossils are usually proposed to be the product of reworking of older sediments that have been eroded and re-deposited in younger sediments. However, fossils *"considered to be reworked do not show any special morphological evidence for having been reworked"* (Woodmorappe, 1982, p. 215). Out-of-place index fossils are thereby arbitrarily explained away without the need for direct physical evidence.

Only by preemptively assuming Biologic Evolution is fact would anyone consider these out-of-place index fossils to have been reworked.

In most cases, the youngest index fossils are used to date the rock layer and the oldest index fossils are proposed to have been reworked. However, in some instances there is a need to use the older index fossils to date the layer in order to preserve Biologic Evolution assumptions. In those instances, a mechanism called *"downwashing"* is imposed. Younger index fossils from clay-rich (argillaceous) layers above are supposedly eroded without significant weathering and presumably washed down through fractures into the older layer below. The fractures supposedly heal without a trace, leaving index fossils of younger age mixed with index fossils of older age in the lower layer. Without supposed reworking and downwashing, the time brackets assigned to index fossils would overlap and quickly discredit their usefulness.

Index fossils are frequently found in the wrong sequence according to the Geologic Column. Woodmorappe (1982, pp. 210-214) has documented 232 stratigraphically out-of-place fossil examples. Supposedly older index fossils positioned above supposedly younger index fossils are most reasonably explained by ecosystem recolonization of species that never became extinct. **Index fossils are exposed as an unreliable Geologic Column construct.**

Dragons Among Us

Paleontologist Sir Richard Owen named the taxon "Dinosauria" in 1842. According to the Geologic Time Scale, dinosaurs became extinct over sixty million years before man's earliest hominoid ancestors walked the Earth. Yet, legends so prevalent around the world, speak of ferocious reptilian beasts called "dragons". Legends are often embellished accounts, rooted in fact. The key to establishing dragon legend credibility is matching descriptions and pictures of these terrible legendary lizards with known dinosaur species. If ancient peoples accurately described and drew creatures that closely resemble actual dinosaurs, they probably observed living dinosaurs.

"A well-known old science book, the Historia Animalium, claims that dragons were still not extinct in the 1500's. But the animals were said to be extremely rare and relatively small by then" (Taylor, P., 1987, p. 41).

Chinese books and art offer thousands of dragon stories and pictures. *"In ancient China, dinosaur bones were thought to be the remains of dead dragons and were ground up for medicine and magic potions"* (Gardom and Milner, 1993, p. 91).

The Catholic Bible (Daniel 14 NAB) speaks of a dragon called "Bel" that was held in captivity by Cyrus the Persian, King of Babylonia during the sixth century B.C. Bel was worshipped by the Babylonians and fed forty sheep, flour, and wine every day. Daniel fed the dragon cakes of pitch, fat, and hair to kill it and to prove that Bel was not an immortal god.

The legend of St. George speaks of him slaying a dreaded dragon in England between 250 and 300 A.D. Similar creatures are commonly portrayed in ancient Chinese drawings. These pictures strongly resemble the dinosaur species Baryonyx that was subsequently discovered in Great Britain in 1983 (Gish, 1992, pp. 80-81).

The city of Nerlus, France, was renamed in honor of killing a dragon there. The dragon was bigger than an ox, with long, sharp, pointed horns on its head (Taylor, P., 1987, p. 40). Horned dinosaurs such as Triceratops are likely candidates.

An Irish writer in 900 A.D. described a large beast with "iron nails" on its tail that pointed backwards. Its head was shaped similar to a horse. It had thick legs and sharp claws. These details strongly resemble a Kentrosaurus or Stegosaurus (Taylor, P. 1987, p. 43).

The Book of Job (Chapter 40) in the Bible refers to a swamp-dwelling vegetarian known as "behemoth" perhaps as recently as the sixth century B.C. Behemoth is described as having bones like *"tubes of bronze"* and a tail that *"sways like a cedar"*. *"When the river rages, he (behemoth) is not alarmed."* This description bears a strong resemblance to Sauropod dinosaurs including Diplodocus, Brachisaurus, or Ultrasaurus.

In the Ndoki region of northern Zaire is a vast area of dense jungle and treacherous swamp. Natives describe an elephant-sized vegetarian creature they call "Mokele-Mbembe". He has smooth, brownish-gray skin, a long flexible neck, a long powerful tail, and leaves three-clawed, frying pan-sized footprints (Anonymous, 1980b, pp. 6-7; Leary, 1980). The natives claim Mokele-Mbembe will kill you if you get too close, but will not eat you. When shown pictures of many different animals, native witnesses picked Apatosaurus as looking most like "Mokele Mbembe" (Gish, 1992, p. 17).

The Book of Job (Chapter 41) in the Bible speaks of a "fire-breathing leviathan" that was at home in the water and left trails in the mud. This mighty creature thrashed the waters like a boiling caldron. Leviathan's skin was of hard, tightly-jointed folds (possibly scales) that protected him from the sword, spear, dart, and javelin. The large size, strong jaws, great teeth, ability to swim and slither across land, and protected back and underside resemble the giant marine lizard Mosasaurus that grew to 50 feet or more.

A German U-boat captain, Georg von Forstner, described a similar Mosasaurus-like animal. *"On July 30, 1915, our U28 torpedoed the British steamer "Iberian" carrying a rich cargo in the North Atlantic. The steamer sank quickly, the bow sticking almost vertically into the air. When it had been down for about twenty-five seconds, there was a violent explosion. A little later, pieces of wreckage, and among them a gigantic sea animal (writhing and struggling wildly), was shot out of the water to a height of 60 to 100-feet. At that moment, I had with me in the conning tower my officers of the watch, the chief engineer, the navigator, and the helmsman. ... It was about 60 feet long, was like a crocodile in shape, and had four limbs with powerful webbed feet, and a long tail tapering to a point"* (Taylor, P., 1987, p. 49; Gish, 1992, p. 86).

The fabled Loch Ness Monster of Scotland is revealed, by a fuzzy underwater picture, to look much like an Elasmosaurus (Gish, 1992, p. 17).

Flying reptiles with a snake-like body and bat-like wings were described in 460 B.C. by the respected Greek explorer Herodotus (1850). They were small in size and of various colors. *"Large numbers would sometimes gather in frankincense trees. When workers wanted to gather the trees valuable juices, they would use smelly smoke to drive the flying reptiles away."* Herodotus reported them as common in Egypt and Arabia. *"Similar animals (three feet long) were also described in India by the geographer Strabo"* (Taylor, P., 1987, p. 44; Strabo, 1967). These descriptions strongly resemble the small flying reptile Rhamphorhynchus.

A reddish-colored flying lizard called "Kongamato" is described by natives as inhabiting the Jiunda Swamp of northern Zimbabwe. Its bare bat-like wings stretch to a wingspan of four to seven feet. Each eye-witness interviewed identified pictures of a Pterodactyl as most like the Kongamato (Taylor, P., 1987, p. 45).

American Indian legends and pictures of the "Thunderbird" are common in the southwestern United States. The "Tombstone Epitaph" (April 26, 1860) reported the story of two cowboys riding in the desert outside of Tombstone, Arizona. They were startled as they looked up and saw an enormous flying creature gliding through the air, feet extended, preparing to land. The horses and men were frantic with fear. The cowboys killed the creature and cut off its wing tip for a trophy. A closer look at the creature on the ground revealed a long slim body with great claws on its feet and the front of its wings. Its eight-foot-long head was similar to an alligator, with a mouth full of teeth, but had large protruding eyes. The creature's wing was a smooth, tough membrane – like a bat's (Gish, 1992, p. 16). This description matches the Quetzalcoatlus fossil found in Big Bend National Park, Texas in 1972. The Quetzalcoatlus fossil had a wingspan of nearly 48 feet, making it the largest of animals known to fly (Gish, 1992, p. 58). David Woetzel (2006) further documents the evidence of historical and possible modern day presence of flying reptiles.

Tyrannosaurus rex bones being examined at Montana State University Lab revealed that some parts deep inside the long leg bone had not completely fossilized. Microscopic examination of a thin section of the bone contained blood vessel channels. *"The lab filled with murmurs of amazement, for I had focused on something inside the vessels ... tiny round objects, translucent red with a dark center. Then a colleague took one look at them and shouted, "You've got red blood cells. You've got red blood cells!"* (Schweitzer and Staedter, 1997 pp. 55-57).

The Liscomb Bone Bed along the Colville River on the North Slope of Alaska contains thousands of frozen, un-fossilized dinosaur bones. An expedition brought back over two hundred pounds of bones to be studied. Some tough tail ligaments are still attached to the bones (Davis, Buddy, 1998, pers. comm.). During a frozen condition, bones and ligaments gradually deteriorate. That implies their age is no more than a few thousand years.

Soft tissue has been discovered in two Tyrannosaurs and a Hadrosaur (duckbill dinosaur), raising the possibility of recovering DNA (Storkstad, 2005, p. 1852). *Research revealed "transparent, flexible, hollow blood vessels containing small round microstructures ... Some regions ... are highly fibrous, and the matrix possesses elasticity and resilience"* (Schweitzer, et al., 2005 p. 1952). The preservation of soft tissue is attributed to dense surrounding bone mineralization *"combined with as yet undetermined geochemical and environmental factors"* (Schweitzer, et al., p. 1955).

Soft tissue is extremely vulnerable to rapid decomposition by scavengers, bacteria, oxidation, heat, water, enzymes, mineralization, and environmental processes (Sonmor, 2005, p. 3). *"Half-life measurements for two molecules in particular – bone collagen and bone mitochondrial DNA – are now rigorously established, and suggest maximum time-to-dust age estimates given assumed temperatures. At reasonable earth surface temperatures, these rates predict near-total molecular degradation on the order of hundreds of thousands of years for collagen, and approximately ten thousand years for DNA. These maximum age estimates are incongruent with fossil age assignments in general, and incongruent by orders of magnitude with Mesozoic original biochemistry discoveries"* (Thomas, Brian, 2013, p. 16).

It is unreasonable to assume that soft tissue could be preserved for millions of years without deteriorating. However, Brian Thomas (2013, p. 3) lists 38 peer-reviewed journal articles documenting fossil soft tissue in sedimentary formations ranging from 10 million to 417 million years according to the Geologic Column time scale. **Fossil and historical evidences support the former and perhaps recent coexistence of dinosaurs and humans.**

Man Before His "Evolutionary" Time

Human presence is confirmed by a fossil human ring finger found between Cretaceous layers near the Paluxy River in Texas. *"This fossil also has the color, texture, and microstructure matching some of the Glen Rose limestone. The joint itself is a distinctive match for the medial joint of the human finger. The fingernail and the cuticle are clearly formed. The end of the fossil tapers, matching the form of a feminine finger. Again, we took the fossil to Cordel Vanhouse for sectioning. ... The interior structure reveals replaced bone marrow, bone shaft, joint flare, flesh proportions, and skin epidermis! All these elements are replaced, but their structure is unmistakable. The fossil is approximately twenty percent longer than the average modern feminine finger"* (Baugh, 1991, p. 169, X).

The fossilized bones of ten humans, collectively known as "Malachite Man" have been excavated fifty feet down into the Cretaceous Dakota Sandstone from open pit copper mine excavation near La Sal, Utah. The fossils appear to have been buried in unconsolidated material which subsequently solidified around them, perhaps during intrusion of copper-rich minerals (Patton, D.R., 6/9/06, Denver presentation).

The Taylor Trail in the Paluxy riverbed near Glen Rose, Texas, reveals what appear to be 14 human tracks in right/left sequence alongside and within dinosaur tracks. The trail led under rock layers which were excavated, extending the trail of fossil footprints (Baugh, 1991, pp. 21-31).

A primitive iron hammer with wooden handle was found contained in Ordovician sandstone near London, Texas (Baugh, 1991, pp. 1-2, Q). This obvious human artifact demonstrates human presence at the time that Ordovician sandstone was deposited. **Evidence of human presence during Paleozoic and Mesozoic sediment deposition discredits Biologic Evolution assumptions regarding the Geologic Column.**

REFERENCES

A.A.B. & J.P., (2012), "And now, the space-weather forcast" (neutrinos and solar storms), *The Economist,* 23 August, pp. 1-2.
<www. economist.com/blogs/babbage/2012/08/neutrinos-and-solar-storms>

Adamenko, S., F. Selleri and A. vander Merwe, (2007), "Controlled Nucleosynthesis- Breakthroughs in Experiment and Theory," Fundamental Theories of Physics, Vol. 156, Springer-Verleg, the Netherlands.

Ager, Derek V., (1993), The New Catastrophism, U.K.: Cambridge University Press, 231 pp.

Aliyev, Ad.A, (2004), "Mud volcanism of the South-Caspian oil-gas basin," *National Academy Sciences of Azerbaijan Institute of Geology,* pp. 186-212).

Allen, David, 1996, "Sediment Transport and the Genesis Flood – Case Studies including The Wawkesbury Sandstone, Sydney," *Creation Ex Nihilo Technical Journal,* Vol. 10, Part 3, p. 362.

Ameglio, L., J.-L. Vigneresse and J.L. Bouchez, (1997), Granite pluton geometry and emplacement mode inferred from combined fabric and gravity data," in
Granite: From Segregation of Melt to Emplacement Fabrics, J.L. Bouchez, D.H.W. Hutton and W.E. Stephens, eds., Dordrecht, The Netherlands: Kluwer Academic Publishers: pp. 199-214.

Ameglio, L. and J.-L. Vigneresse, (1999), "Geophysical imaging of the shape of granitic intrusions at depth": A review, in *Understanding Granites: Integrating New and Classical Techniques,* A. Castro, C. Fernandez, and J.-L. Vigneresse, eds., London: The Geological Society, Special Publication 168: pp. 39-54.

American Journal of the United States, (1971), Vol. 6, p. 210.

Ammon, C.J., et al., (2005), "Rupture process of the 2004 Sumatra-Andaman Earthquake," *Science,* Vol. 308, (May 20), pp. 1133-1139.

Anfiloff, V., (1992), "The tectonic framework of Australia," in: Chatterjee, S. and N. Hotton III (editors), New Concepts in Global Tectonics, Texas Tech University Press, Lubbock, TX, pp. 75-109.

Anonymous, (1980a), "Magsat down, magnetic field declining," *Science News*, Vol. 117, (28 June), p. 407.

Anonymous, (1980b), "Living dinosaurs," *Science*, Vol. 1, (November), pp. 6-7.

Armitage, Mark, (2001), "New record of polonium radiohalos, Stone Mountain Granite, Georgia (USA), *TJ*, Vol. 15 (1), pp. 86-88.

Austin, Steven A., (1990), "Were Grand Canyon limestones Deposited by calm and placid seas?," Institute for Creation Research, El Cajon, CA, (December), Impact Article No. 210, p. 1.

Austin, S.A., J.R. Baumgardner, D.R. Humphreys, A.A. Snelling, L. Vardiman, and K.P. Wise, (1994), "Catastrophic plate tectonics: a global flood model of Earth history," in: Walsh, R.E. (editor), Proceedings of the Third International Conference on Creationism, Creation Science Fellowship, Pittsburgh, PA, pp. 609-621.

Austin, Steven A. and Mark L. Strauss, (1999), "Are earthquakes signs of the end times," *Christian Research Journal,* Vol. 21, (4), pp. 30-39.

Austin, Steven A., (2010), "Supervolcanoes and the Mount St. Helens eruptions," *Acts And Facts,* 39(5): p. 249.

Awramik, Stanley M. and Kathleen Grey, (2005), "Stromatolites: Biogenicity, Biosignatures, and Bioconfusion", Dept. of Geological Sciences, University of California, Santa Barbara, CA 93106, USA; awramik@geol.ucsb.edu, pp. 1-9.

Baker, E.R., 1996, "Granitic melt velocities: Empirical and configurational entropy models for their calculation," *Americal Mineralogist,* Vol 81., pp. 126-134).

Baker, V., G. Benito, and A. Rudoy, (1993), "Paleohydrology of Late Pleistocene Superflooding, Altay Mountains, Siberia," *Science*, Vol. 259, (15 January), pp. 348-350.

Barenbaum, A.A., (2004), Mechanisms of formation of oil and gas deposits, *Proceedings of Russian Academy of Science,* Vol. 399, No. 6, pp. 802-805.

Baugh, Carl E. with C.A. Wilson, (1991), Dinosaur, Promise Publishing Co., Orange, CA 92667, 170 pp.

Baumgartner, J. R., A. A. Snelling, D. R. Humphreys, and S. A. Austin, (2003), "Measurable C-14 in fossilized organic materials confirming the Young Earth Creation- Flood Model," in Proceedings of the Fifth International Conference on Creationism 2003, pp. 127-142.

Beloussov, V.V., (1980), *Geotectonics,* Moscow: Mir.

Bengtson, S., V. Belivanova, B. Rasmussen, M. Whitehouse, (2009), "The controversial "Cambrian" fossils of the Vindhyan are real but more than a billion years older" 6 pp.

Bergeron, Lou, (1997), "Deep Waters," *New Scientist*, Vol. 155, No. 2097, (30 August), pp. 22-26.

Berthault, Guy, (1997), "Sedimentation experiments: Is extrapolation Appropriate? A reply," *Creation Ex Nihilo Technical Journal,* Vol. 11, Part 1, pp. 65-70.

Bloom, A.L., (1998), *Geomorphology (3rd ed.),* Waveland Press, Long Grove IL.

Bloxham, Jeremy and D. Gubbins, (1989), "The evolution of the Earth's magnetic field," *Scientific American*, Vol. 261, December, pp. 68-75.

Boggs, Sam Jr., (1986), Sedimentary Environments and Facies, 2nd Edition, p. 193.

Boudreaux, Edward A., (2005-2013), personal communication; 2910 E. 121st Court; Thornton, CO 80241 or eaboudre@yahoo.com.

Boudreaux, Dr. Edward A. and Eric C. Baxter, (2010), *God Created the Earth – Genesis of Creation Chemistry*, Central Ohio Creation Research Association; 1007 Groveport Road; Canal Winchester, OH 23110, 94 pp.

Boudreaux, Dr. Edward A. and Eric C. Baxter, (2012), *God Created the Earth – Genesis of Creation Chemistry*, 2nd Edition, Rocky Mountain Creation Fellowship; P.O. Box 3451; Littleton, CO 80161; 122 pp.).

Bracken, Rob, (2012), pers. comm., P.O. Box 1408; Golden, CO 80402.

Broadhead, G.C., (1902), "The New Madrid Earthquake," *American Geologist*, Vol. 30, No.2, (August), pp.76-87.

Brown, Walter T. Jr., (1989), In the Beginning, Fifth edition, Center for Scientific Creation, Phoenix, AZ, 122 pp.

Brown, Walter T. Jr., (2001), *In the Beginning; compelling Evidence for creation and the Flood*, Seventh Edition, Center for Scientific Creation; 5612 N. 20th Place; Phoenix, AZ 85016, 328 pp.

Bruce, Richard, (2000), pers. comm., Mineral House; 100 Plain Street; Western Australia 6004.

Buckland, W., (1937), *Geology and Mineralogy*, Philadelphia, p. 101.

Budge, Sir E. A. W., editor, (1927), *The Book of the Cave Treasures*, Translated from the Syriac Text (written about 300-599 A.D.) of the British Museum (MS. Add. 25875), The Religious Tract Society, London, England.

Burnham, C.W., (1979), *Geochemistry of Hydrothermal Ore Deposits*, 2nd Edition, Barnes, H.L. (editor), Wiley, New York, pp. 71-136.

Byers, Horace Robert, (1974), *General Meterology*, 4th ed., McGraw-Hill, New York.

Cardone, F. and R. Mignani (2007), *Deformed Spacetime*, Chps. 16-17, Springer, Dordrecht.

Cardone, F., A. Carpinteri and G. Lacidogna, (2009), "Piezonuclear neutrons from Fracturing of inert solids," *Physics Letters A,* Vol. 373, pp. 4158-4163.

Cardone, F., G. Cherubini and A Petrucci, (2009), "Piezonuclear neutrons," *Physics Letters A,* Vol. 373, pp. 862-866.

Carpinteri, A, G. Lacidogna, A. Manuello and O. Borla, (2011), "Piezonuclear fission reactions in rocks: Evidences from microchemical analysis, neutron emission, and geological transformation, *Rock Mechanics and Rock Engineering,* Springer Wien, New York, 17 pp.

Cassadevall, T.J. and L.P. Greenland, (1980), "The chemistry of gases emanating from Mount St. Helens," *U.S. Geological Survey Professional Paper 1250*, pp. 221-226.

Cengage, Gale, (2003), *World of Earth Sciences.*

Chamberlin, R.T., (1937), "The origin and history of the Earth," in: Moulton, F.R. (editor), *The World and Man*.

Chatterjee, S. and N. Hotton III (editors), (1992), *New Concepts in Global Tectonics,* Texas Tech University Press, Lubbock, TX, 450 pp.

Chien, Paul, (1997), "Explosion of Life," interview, 30 June, Biology Department Chairman, University of San Francisco.

Chopin, C., (1987), "Very high-pressure metamorphism in the Western Alps: implications for subduction of continental crust," in: Oxburgh, E.R. (editor), Tectonic Settings of Regional Metamorphism: Proceedings of a Royal Society Discussion Meeting, Royal Society of London, London, pp. 183-197.

Clemens, J.D. and C.K. Mawer, (1992), "Granitic magma transport by fracture propogation," *Tectonophysics,* Vol. 204, pp. 339-360.
Clemens, J.D., N. Petford, and C.K. Mawer, (1997), "Ascent mechanisms of granitic magmas: Causes and consequence," in *Deformation-Enhanced Fluid Transport in the Earth's Crust and Mantle,* M. Holness, ed., London: Chapman and Hall, pp. 145-172.
Clemens, J.D. and N. Petford, (1999), "Granitic melt viscosity and silicic magma dynamics in contrasting tectonic settings," London: *Journal of the Geological Society,* Vol. 156, pp. 1057-1060.)
Clemens, J.D., (2005), "Granites and granitic magmas: strange phenomena and new Perspectives on some old problems, *Proceedings of the Geologists' Association,* Vol. 116, pp. 9-16.
Collins, Lorence G., (1997), Contrasting Characteristics of Magmatic and Metasomatic Granite and the Myth that Granite Plutons can be only Magmatic.
Cousin, R., G. Brenton, and J.P. Watt, (1994), "Dinosaur egglaying and nesting in France," in: Carpenter, K., K.F. Hirsch, and J.R. Horner (ed.), *Dinosaur Eggs and Babies*, Cambridge University Press, London, pp. 56-74.
Creation Magazine, (1990), 13(1), pp. 10-14.
Crickmay, C.H, (1975), "The hypothesis of unequal activity" in: Welhorn, W.N., and Flemal, R.C. (Eds.), *Theories of Landform Development*, George Allen and Unwin, London, pp. 107-108.
Currie, P.J., G.C. Nandon, and M.G. Lockley, (1991), "Dinosaur footprints with skin impressions from the Cretaceous of Alberta and Colorado," *Canadian Journal of Earth Sciences*, No. 28, pp. 102-115.
Dana, Edward S. and James D. Dana, (1959), *Dana's Manual of Mineralogy*, 17th Edition, John Wiley &Sons, Inc., 609 pp.
Darwin, Charles, (1835), *Geological Observations on the Volcanic Islands and Parts of South America*, Part II, Chapter 15.
Dashzeveg, D., M.J. Novacek, M.A. Norell, J.M. Clark, L.M. Chiappe, A. Davidson, M.C. McKenna, L. Dingus, C. Swisher, and P. Altangerel, (1995), "Extraordinary preservation in a new vertebrate assemblage from the Late Cretaceous of Mongolia," *Nature*, No. 374, pp. 446-449.
Davis, Buddy (1998-2000), (pers. comm.), Answers in Genesis; P.O. Box 6330; Florence, KY 41022-6330.
Derry, D.R., G.R. Clark, and N. Gillatt, (1965), "The Northgate base metal deposit at Tynagh Co. Galway, Ireland," *Economic Geology*, No. 60, pp. 1218-1237.
deTerra, Helmut and T.T. Paterson, (1939), *Studies on the Ice Age in India and Associated Human Cultures*, pp. 222-225.
Dethier, D., D. Pevear, and D. Frank, (1981), "Alteration of New Volcanic Deposits, " *The 1980 eruptions of Mount St. Helens, Washington,* U.S. Geologic Survey Professional Paper 1250, p. 663.
DeYoung, Donald B., (1995), Astronomy and the Bible, Baker Books; P.O. Box 6287; Grand Rapids, MI, 146 pp.
Dikin, A.P., (1997), *Radiogeneic Isotope Geology,* (updated paperback edition), U.K., New York: Cambridge University Press.

Dillon, L.S., (1974), "Neovolcanism: a proposed replacement for the concepts of plate tectonics and continental drift," in: Kahle, C.F. (ed.), *Plate Tectonics Assessments and Reassessments*, American Association of Petroleum Geologists Memoir 23, Tulsa, OK, pp. 167- 239.

Dillow, Joseph, (1982), *The Waters Above: Earth's Preflood Vapor Canopy*, Moody Press, Chicago, IL., 479 pp.

Dmitrievsky, D. and B.M. Valyaev (editors), (2002), *Degassing of the Earth: Geodynamics, Geofluids, Petrol and Gas,* GEOS Publishers, Moscow (in Russian).

Dohrenwend, J.C., (1994), "Pediments in arid environments," in: Abrahams, A.D. and A.J. Parsons (editors), *Geomorphology of Desert Environments*, Chapman and Hall, London, p. 321.

Doughty, J.R., (2006), "Isotopic analysis of Fruitland Formation coal bed carbon dioxide and methane," *Creation Research Society Quarterly,* Vol. 43, No. 2, pp. 105-110.

Einsele, Gerhard, (2000), *Sedimentary basins: evolution, facies, and sediment budget,* 792 pp.).

Emery, G.T., (1972), "Perturbations of Nuclear Decay Rates," *Annular Review of Nuclear Science,* Vol. 22, p. 165.

Erman, G.A., (1848), *Travels in Siberia*, No. II, pp. 376, 383.

Evans, D.J., W.J. Rowley, R.A. Chadwick, E.S. Kimbell and D. Millward, (1994), "Seismic reflection data and the internal structure of the Lake District batholith, Cumbria, northern England," *Proceedings of the Yorkshire Geological Society,* Vol. 50, pp. 11-24.

Evans, David M., (1966), "Man-made earthquakes in Denver," *Geotimes*, May/June, pp. 11-18.

Ewing, Maurice, (1948), "Exploring the Mid-Atlantic Ridge," *National Geographic,* Vol. XCIV, No. 3, September. pp. 275-294.

Ewing, Maurice, (1949) "New discoveries on the Mid-Atlantic Ridge," *National Geographic Magazine*, Vol. XCVI, No. 5, pp. 611-640.

Ezra, (no date), *II Esdras*, 6:41, 42, (extra-biblical writing).

Facca, G. and F. Tonani, (1964), "Theory and technology of a geotghermal field," *Bulletin Volcanology,* No. 27, pp. 1-47.

Farquhar, Brodie, (2007), "Yellowstone hot spot bulges, perplexes scientists," *Yellowstone Journal*, Vol. 14, Ed. 1, June, pp. YJ-12,13.

Ferrell, Vance, (2005), Evolution Handbook, Evolution Facts, Inc.; Box 300; Altamont, TN 37301, evolution-facts.org, 992 pp.

Firstbrook, P., (1999), *Lost on Everest: the search for Mallory and Irvine*, BBC Worldwide, London.

Folk, R.L., (1959), "Practical petrographic classification of limestones," *American Association of Petroleum Geologists Bulletin,* No. 43, p. 8.

Froede, Carl R. Jr. and A. Jerry Akridge, (2013), "Authors' Response." *Creation Research Society Quarterly,* Vol. 49, No. 4, pp. 332-334.

Gale, N. et al., (1972), "Uranium-lead chronology of chondrite meteorites," *Nature* (Physical Sciences), Vol. 240, 20 November, p. 57.

Gallagher, Ronnie, (2003), "Mud Volcanoes mysterious phenomena fascinate scientists and tourists," *Azerbaijan International*, Summer (11.2), pp. 44-49.

Gansser, A., (1964), *Geology of the Himalayas*, Interscience Publishers, New York.

Gardom, Tim and Angela Milner, (1993), *The Natural History Book of Dinosaurs*, Carlton Books Limited, 128 pp.

Garner, Paul, (1996), "Where is the Flood/post-Flood boundary? Implications of dinosaur nests in the Mesozoic," *Creation Ex Nihilo Technical Journal*, Vol. 10, Part 1, Answers in Genesis; P.O. Box 6330; Florence, KY 41022-6330, pp. 101-106.

Garton, Michael, (1996), "The pattern of fossil tracks in the Geologic Record," *Creation Ex Nihilo Technical Journal*, Vol. 10, Part 1, Answers in Genesis; P.O. Box 6330: Florence, KY 41022-6330, pp. 82-100.

Gentry, Robert V., (1986), *Creation's Tiny Mystery*, Earth Science Associates, Knoxville, TN 37912-0067, 315 pp.

Gentry, Robert V., (1988), *Creation's Tiny Mystery*, 2nd edition, Earth Science Associates, Knoxville, TN 37912-0067, 347 pp.

Gentry, Robert V., (1990), "Critique of radiohalo evidence regarding change in natural process rates," *Creation Research Society Quarterly*, Vol. 27, No. 3, December, pp. 103-105.

Gidley, (1918), in: *Explorations and Fieldwork of the Smithsonian Institution for the year 1913*, Washington, 1914; Annual Report for the Smithsonian Institution for 1918, pp. 281-287.

Gish, Duane T., (1992), *Dinosaurs by Design*, Master Books; Green Forest, AR, 88 pp.

Glassley, W.E., (1983), "The role of CO-2 in the chemical modification of deep continental crust," *Geochim, Cosmochim, Acta*, No. 47, pp. 597-616.

Goddard, A., (1995), "Burst zip undermines theories of ocean birth," *New Scientist*, No. 146, (1723), p. 14.

Gold, T. and S. Soter, (1980), The deep-earth-gas hypothesis," *Scientific American*, Vol. 242, No. 6, pp. 154-161.

Gollmer, Steven M., (2013), *"Initial conditions for a post-Flood Rapid Ice Age,"* Proceedings of the Seventh International Conference on Creationism, Creation Science Fellowship, Pittsburgh, PA, USA.

Gore, D.B. and M.P. Taylor, (2003), Discussion and reply –Grooves and striations on the Stanthorpe Adamellite; Evidenced for a possible late Middle – late Triassic age glaciation discussion, *Australian J. Earth Sciences*, Vol. 50, pp. 467-470.

Gregory, (1896), "The Great Rift Valley," *Geographical Journal*, No. IV (1894), pp. 5, 6, 236.

Gretener, P.E., (1972), "Thoughts on overthrust faulting in a layered sequence," *CA* 20:590-1.

Gretener, P.E., (1977), "On the character of thrust faults with particular reference to the basal tongues," *CA*, 25:111.

Gross, Richard (2010), "Chile quake moved Earth's axis, shortened days," *NASA Jet Propulsion Laboratory, California*, 2 March.

Guth, Peter L., K.V. Hodges, and J.H. Willemin, (1982), "Limitations on the role of pore pressure in gravity gliding," *Geological Society of America Bulletin*, Vol. 93 July, pp. 606-612.

Habermehl, Anne, (2013), "Ancient Egypt, the Ice Age, and Biblical chronology," *Proceedings of the Seventh International Conference on Creationism*, Creation Science Fellowship, Pittsburgh, PA, USA.

Hadley, R.F., (1967), "Pediments and pediment-forming Processes, *Journal of Geological Education*, Vol. 15, No. 83, pp. 83-89.

Hall, J.M. and P.T. Robinson, (1979), "Deep crustal drilling in the North Atlantic Ocean," *Science*, Vol. 204, 11 May, pp. 573-586.

Hamilton, W.B., (2003), "An Alternative Earth," *GSA Today*, Vol. 13, No. 11, pp. 4-12.

Hancock, Graham, (2002), *Underworld*, Crown Publishing, New York, NY, p. 477.

Hansen, E.C., R.C. Newton, and A.S. Janardhan, (1984), "Fluid inclusions in rocks from the amphibolite- facies gneiss to charnockite progression in southern Karnataka, India: direct evidence concerning the fluids of granulite metamorphism," *Journal Metamorphic Geology*, No. 2, pp. 249-264.

Hapgood, Charles H., (1996), *Maps of the Ancient Sea Kings*, Adventures Unlimited Press; One Adventure Place; Kempton, IL 60946, USA, 315 pp.)

Heckel, P.H., (1974), "Carbonate buildup in the geologic record: A review, in *Reefs in Time and Space*, L.F. Laporte, ed., Society of Economic Paleontologists and Mineralogists Special Publication 18; pp. 90-154.

Heim, Arnold and August Gausser, (1939), *The Throne of the Gods, an account of the first Swiss expedition to the Himalayas*, p. 218.

Herodotus, (1850), *Historiae*, Henry Clay (translator), Henry G. Bohn, London, UK.

Hibben, Frank C., (1943), "Evidence of early man in Alaska," *American Antiquity*, No. VIII, p. 256.

Hibben, Frank C., (1951), *Treasure in the Dust*, p. 56.

Holroyd, Edmond W. III, (1999-2011), (pers. comm.), 8905 W. 63rd Avenue; Arvada, CO 80004-3103.

Holt, Roy D., (1996), "Evidence for a Late Cainozoic Flood/ post-Flood boundary," *Creation ex Nihilo Technical Journal*, Vol. 10, Part 1, Answers in Genesis; P.O. Box 6330; Florence, KY 41022-6330, pp. 128-163.

Hong, Z., et al., (1990), "Early Archean inheritance in zircon from Mesozoic Dalngshan granitoids in the Yangtze Foldbelt of southeast China, *Geochemical Journal (of Japan)*, Vol 24, pp. 133-141.

Horner, J.R. and J. Gorman, (1988), *Digging Dinosaurs*, Workman Publishing, New York.

Hovland, M. and A.G. Judd, (1988*), Seabed Pockmarks and Seepages – Impact on Geology, Biology, and the Marine Environment*, Graham and Trotman, London, pp. 58-118.

Hovland, M., H.G. Rueslatten, H.K. Johnsen, B. Kvamme, and T. Kuznetsova, (2006), "Salt formation associated with sub-surface boiling and supercritical water," *Marine and Petroleum Geology*, Vol. 23, Issue 8, pp. 855-869.

Howorth, H.H., (1880), "The Mammoth in Siberia", *The Geological Magazine*, September, pp. 408-414.

Hsu, Kenneth J., (1991), "Exhumation of high-pressure metamorphic rocks," *Geology*, No. 19, pp. 107-110.

Hubbert, M.K., and W.W. Rubey, (1959), "Role of fluid pressure in the mechanics of overthrust faulting," *Geologic Society of America Bulletin*, Vol. 70, February, pp. 115-166.

Ivester, A.H., D.I. Godfrey-Smith, J.J. Brooks, and B.E. Taylor, (2004), "The timing of Carolina Bays and inland activity on the Atlantic Coastal plain of Georgia and South Carolina," *Geological Society of America Abstracts with Programs*, Vol. 36, No. 5, p. 69).

Janaardhan, A.S., R.C. Newton, and E.C. Hansen (1982), "The transformation of amphibolite facies gneiss to charnockite in southern Karnataka and northern Tamil Nadu, India," *Contributions Mineral. Petrol.*, No. 79, pp. 130-149.

Johnson, D. (1931), "Planes of lateral corrosion," *Science*, Vol. 73, p. 174.

Johnson, Douglas, (1942), *The Origin of the Carolina Bays*.

Johnstone, M.H., P.J. Jones, W.J. Koop, J. Roberts, J. Gilbert-Tomlinson, J.J. Veevers, and A.T. Wells, (1967), "Devonian of Western and Central Australia," in: Oswald, D.H. (editor), *International Symposium Devonian System, Calgary*, No. 1, pp. 599-612.

Kaczorowski, R.T., (1977), *The Carolina Bays: a comparison with Modern Oriented Lakes*, Technical Report no. 13-CDR, Coastal research Division, Department of Geology, University of South Carolina, Columbia, South Carolina, 124 pp.

Kadirov, F.A., I. Lerche, L.S. Guliyev, A.Sh. Mukhtarov, A.H. Kadyrov, A.A. Feyzullayev, and Ch.S. Aliyev, (2004), "Mud Volcanoes: Deep structures, dynamics and post-explosion thermal conditions," *National Academy Sciences of Azerbaijan Institute of Geology*, pp. 223-256.

Keith, M.L., (1993), Geodynamics and mantle flow: an alternative earth model, *Earth-Science Reviews,* vol. 33, pp. 153-337.

Kelly, J.T., S.M. Dickson, D.F. Belknap, W.A. Barnhart, and M. Henderson, (1994), "Giant sea-bed pockmarks: Evidence for gas escape from Belfast Bay, Maine," *Geology*, No. 22 (1), pp. 59-62.

Kerr, Richard A., (1987), "Ocean hot springs similar around the globe," *Science*, Vol. 235, p. 435.

Kerr, Richard A., (1989), "Deep holes yielding geoscience suprises," *Science*, Vol. 245, pp. 468-470.

King, L.H. and B. MacLean, (1970), "Pockmarks on the Scotian Shelf," *Geological Society of America Bulletin*, No. 81, pp. 3141-3148.

Kircher, (1828), *American Journal of Science and Arts*, Vol. 14, p. 75.

Knapp, C.C., J.H. Knapp, and J.A. Connor, (2004), "Crustal-scale structure of the South Caspian Basin revealed by deep seismic reflection profiling," *Marine and Petroleum Geology,* 21, pp. 1073-1081.

Kolbe, E., K.H. Langanke, and G.M. Fuller. (2004), *Physical Review Letters,* Vol. 92, Issue 11, 4pp.

Kolbe, E., K.H. Langanke, and G.M. Fuller, (2008), "Neutrino-induced fission of neutron-rich nuclei," 2 February, 5 pp. (from Physics Letters B, Volume 616, Issue 1-2, pp. 48-58.)

Kolbe, R.W., (1957), "Fresh Water Diatoms from Atlantic Deep-Sea Sediments," *Science,* Vol. 126, No. 3282, 22 November.

Kolbe, R.W., (1958), "Turbidity Currents and Displaced Fresh-Water Diatoms," *Science,* Vol. 127, No. 3313, 27 June, pp. 1504-1505.

Krauskopf, K.B., (1967), *Introduction to Geochemistry*, McGraw Hill.

Krumbein, W.C. and L.L. Sloss, (1963), *Stratigraphy and Sedimentation*, Second Edition, W.H. Freeman and Company, San Francisco.

Kunk, M.J. and J.F. Sutter, (1984), "40-Ar/39-Ar age spectrum dating of biotite from Middle Ordovician bentonites – eastern North America, pp. 11-22, in D.L. Bruton, ed., *Aspects of the Ordovician System,* University of Oslo Press, 228 p.

Lalomov, Alexander V., (2007), "Mineral Deposits as an Example of Geological Rates", *Creation Research Society Quarterly,* Vol. 44, No. 1, pp. 64-66.

Leary, Warren E., (1980), "Dinosaurs may inhabit remote jungle," *San Diego Tribune*, 18 October, (Washington Date Line).

Lee, J.S., (1939), *The Geology of China*, London, p. 370.

Lingenfelter, Richard E., (1963), "Production of Carbon-14 by cosmic ray neutrons," *Reviews of Geophysics*, Vol. 1, No. 1, February, p. 51.

Lucchitta, Ivo, (1990), "History of the Grand Canyon and of the Colorado River in Arizona," in *Grand Canyon Geology,* S.S. Beus and M. Morales eds., Oxford University Press, 200 Madison Avenue, New York, New York 10016, 518 pp.

Lumpkin, Joseph B., (2010), *The Encyclopedia of Lost and Rejected Scriptures: The Pseudepigrapha and Apocrypha,* Fifth Estate Publishers; P.O. Box 1116; Blountsville, AL 35031, 825 pp.

Luyendyk, B.P. and W.G. Melson, (1967), "Magnetic properties and petrology of rocks near the crest of the Mid-Atlantic Ridge," *Nature,* Vol. 215, No. 5097, pp. 147-149.

Lynch, Brendan M., (2011), "Research overturns oldest evidence of life on Earth, with implications for Mars, *University of Kansas* 785-864-8855, March 15, p. 1).

Macgowan, K., (1950), *Early Man in the New World*, p. 151.

Mammerickx, J., (1964), "Quantitative observations on pediments in the Mojave and Sonoran Deserts (South-western United States)," *American Journal of Science*, Vol. 262, pp. 417-435.

Markham, Clemens, (1910), *The Incas of Peru.*

Marshall, Craig P., J.R. Emry, and A.O. Marshall, (2011), "Haematite pseudomicrofossils present in the 3.5- billion-year-old Apex Chert", *Nature Geoscience,* 20 February DOI:10.1038/NGEO1084, Macmillian Publishers Limited, pp. 1-4.

Matthews, Mark Z., (2003), "The Oklo Natural Reactors – Evidence of Variable Constants?" *Proceedings of the Fifth International Conference on Creationism,* Creation Science Fellowship, Inc.; 705 Washington Drive; Pittsburg, PA 15229; pp. 209-217.

McDougall, I., et al., (1969), "Excess radiogenic argon in young subaerial basalts from Auckland Volcanic Field, New Zealand," *Geochemica et Cosmochemica Acta,* Vol. 33, pp. 1485-1520.

McKee, E.D., R.F. Wilson, W.J. Breed, and C.S. Breed, (1967), "Evolution of the Colorado River in Arizona," *Museum of Northern Arizona Bulletin 44,* 67 pp.

McKee, E.D. and R.G. Gutschick, (1969), "History of the Redwall Limestone in Northern Arizona," *Geological Society of America Memoir 114*, Boulder, CO, p. 103.

Melson, William G. and Clifford A. Hopson, (1981), "Preeruption Temperatures and Oxygen Fugacities in the 1980 Eruptive Sequence," *The 1980 Eruptions of Mount St. Helens, Washington,* Geological Survey Professional Paper 1250, pp. 641-648).

Meyerhoff, A.A. and H.A. Meyerhoff, (1974), "Ocean magnetic anomalies and their relations to continents," American Association of Petroleum Geologists, Memoir 23, pp. 411-422).

Meyerhoff, A.A., W.B. Agocs, Irfan Tanner, A.E.I. Morris, and B.D. Martin, (1992), "Origin of midocean ridges," *New Concepts in Global Tectonics,* Texas Tech University Press, Lubbock, TX, pp. 151-178.

Miller, Hugh, (1865), *The Old Red Sandstone,* Boston, (first published in England in 1841).

Miller, J.P. & R. Scholten, (1966), "Oceans, Lakes, and Shoreline Features," *Laboratory Studies in Geology,* No. 225.

Monastersky, Richard, (1994), "Great earthquake shakes off Theories," *Science News,* Vol. 154, 9 April, p. 155.

Monroe, J.S. and R. Wicander, (1994), *The Changing Earth: exploring geology and evolution,* West Publishing Company; 610 Opperman Drive; P.O. Box 64526; St. Paul, MN 55164-0526, 731 pp.

Moon, H.P., (1939), "The geology and physiography of the Altiplano of Peru and Bolivia," *The Transactions of the Linnean Society of London,* 3rd Series, Vol. I, Pt. 1, p. 32.

Moore, R., (1953), *Man, Time, and Fossils,* pp. 274-275.

Morris, John D., (1994), *The Young Earth,* Master Books, 141 pp.

Morris, John D., (2003), "Don't the fossils prove Evolution?" *Acts & Facts,* Institute for Creation Research, P.O. Box 2667, El Cajon, CA 92021, Vol. 32, No. 4, April, p. d.

Nadon, G.D., (1993), "The association of anastomosed fluvial deposits and dinosaur tracks, eggs, and nests: implications for the interpretation of floodplain environments and a possible survival strategy for ornithopods," *Palaios,* No. 8, pp. 31-44.

Nienhuis, James I., (2006), *Ice Age Civilizations,* Genesis Veracity; P.O. Box 850; 5773 Woodway Drive; Houston, TX 77057, 1-866-GEOFACT, 200 pp.

Noorbergen, Rene, (1992), *Secrets of the Lost Races: New Discoveries of Advanced Technology in Ancient Civilizations,* Collegedale, TN, Norcom Publishing Corp., p. 3.

Nutting, David, (1994-2006), pers. com., Alpha Omega Institute; P.O. Box 4343; Grand Junction, CO 81502.

Oard, Michael J., (1990), *An Ice Age caused by the Genesis Flood,* Institute for Creation Research.

Oard, Michael J., (1997), *Ancient Ice Ages or Gigantic Submarine Landslides?,* Creation Research Society Monograph No. 6, Creation Research Society; 6801 N. Highway 89; Chino Valley, AZ 86323-9186.

Oard, Michael J., (2002), "The Mountains Rose", (a book review of 'The Origin of Mountains'), *TJ,* Vol. 16, No. 3, pp. 40-43.

Oard, Michael J., (2004), "Pediments formed by the Flood: Evidence for the Flood/post-Flood boundary in the Late Cenozoic," *TJ,* Vol. 18, No. 2, pp. 15-27.

Oard, Michael J., (2004), Frozen in Time; Master Books Inc. P.O. Box 726; Green Forest, AR 72638; 217 pp.

Oard, Michael J., (2005), The Frozen Record: Examining the Ice Core History of the Greenland and Antarctic Ice Sheets, Institute for Creation Research, 199 pp.

Oard, Michael J., (2008), "How did 90% of large Australian Ice Age animals go extinct?" *Journal of Creation,* Vol. 22, No. 1, pp. 17-19.

Oard, Michael J., (2013), "The meaning of porous dinosaur eggs laid on flat bedding planes," *Journal of Creation,* Vol. 27, No. 1, pp. 3-4.

Oard, Michael J., (2013 b), "Raindrop imprints and the location of the pre-Flood/Flood boundary," *Journal of Creation,* Vol. 27, No. 2, pp. 7-8).

Ollier, Cliff and Collin Pain, (2000), *The Origin of Mountains,* (first published 2000 by Routledge; 11 New Fetter Lane, London EC4P 4EE) University Press, Cambridge, United Kingdom, 345 pp.

Ottonello, G., (1997), *Principles of Geochemistry,* Columbia University Press, pp. 145, 487.

Overman, Richard L., (2013), "The temporal, geographical, and geological ubiquity of excess argon with a young-earth analysis," *Proceedings of the Seventh International Conference on Creationism,* Creation Science Fellowship, Pittsburgh, PA, USA.

Paige, S., (1912), "Rock-cut surfaces in the desert ranges, *Journal of Geology*, Vol. 20, p. 444.

Patton, D.R., (2006), June 9th Denver presentation , www.bible.ca/tracks.

Paull, C.K., W. Ussler III, W.S. Borowski, and F.N. Spiess, (1995), "Methane-rich plumes on the Carolina continental rise: associations with gas hydrates," *Geology*, No. 23(1), pp. 89-92.

Perrett, F.A., (1935), *The Eruption of Mt. Pelee 1929-1932,* Carnegie Institution of Washington, pp. 76, 79.

Petford, N., R.C. Kerr, and J.R. Lister, (1993), "Dike transport of granitoid magmas," *Geology,* Vol. 21, pp. 845-848.

Petford, N. and J.D. Clemens, (2000), "Granites are not Diapritic!," *Geology Today,* 16 (5) pp. 180-184.

Pettijohn, F.J., (1975), *Sedimentary Rocks,* Third Edition, Harper and Row, New York.

Polyak, V.J. and P.P. Provencio, (2000), "Summary of the timing of sulfuric-acid speleogenesis for Guadalupe Caves based on ages of alunite," *Journal of Cave and Karst Studies,* Vol. 62, No. 2, pp. 72-79.

Posnansky, A., (1945), *Tiahuanacu, the Cradle of the American Man.*

Pratt, D., (2001), Problems with Plate Tectonics, *New Concepts in Global Tectonics Newsletter,* No. 21, December, pp. 5-24.

Prestwich, (1895), *On Certain Phenomena Belonging to the close of the Last Geological Period and on their bearing upon the Tradition of the Flood*, Macmillian and Co., London

Prouty, W.F., (1952), "Carolina Bays and their origin," *Bulletin of the Geological Society of America*, LXIII, pp. 167-224.

Rainey, F., (1940), "Archaeological investigation in Central Alaska," *American Antiquity*, V, p. 305.

Raymond, Chris, (1991), "Discovery of leaves in Antarctica sparks debate over whether region had near-temperate climate," *Chronicle of Higher Education*, 20 March, pp. A-9, A-11.

Reed, John K., (2000), *Plate Tectonics: A Different View*, Creation Research Society Monograph Series No. 10, Creation Research Society Books; P.O. Box 8263; St. Joseph, MO 64508-8263, 191 pp.

Reed, John K., (2005), "The Geology of the Timbered Hills Group in Oklahoma," *Creation Research Society Quarterly,* Vol. 42, No. 1, June, pp. 39-67.

Rezvoy, P.P., (1971), "Problem of thrust sheets in the Alay Range," *IG*, 13:1736.

Riding, R. (1994), Evolution of algal and cyanobacterial calcification, *Early life on Earth Nobel Symposium 84,* Columbia Univ. Press, New York, NY, pp. 426-438.

Riggs, (1984), *Science*, Vol. 224, April 6, pp. 58-61.

Robinson, Andrew, (1993), *Earth Shock*, Thames and Hudson Ltd., New York, 304 pp.

Robinson, Steven J., (1996), "Can Flood geology explain the Fossil Record?," *Creation Ex Nihilo Technical Journal*, Vol. 10, No. 1, Answers in Genesis; P.O. Box 6330; Florence, KY 41022-6330, pp. 32-69.

Robinson, Steven J., (2003), "Flood/post-Flood boundaries within the global stratigraphical record – Steven Robinson replies:" *TJ*, Vol. 17, No. 3, pp. 53-55.

Robock, A., (2009), "Did the Tuba volcanic eruption of ~74ka B.P. produce widespread Glaciation?," *Journal of Geophysical Research,* 114:D10107.

Roddick, J.C. and M.L. Bevier, (1995), U-Pb dating of granites with inherited zircon, *Chemical Geology,* Vol. 119, pp. 307-329.

Russell, A.J. and O. Knudsen, (1999), "An ice-contact rhythmite (turbidite) succession deposited during November 1996 catastrophic outburst flood (jokulhlaup) Skeidararjokull, Iceland," *Sedimentary Geology*, Vol. 127, pp. 1-10.

Scarborough, R.B., (1989), "Cenozoic Erosion and sedimentation in Arizona," in Jenney, J.P. and S.J. Reynolds, eds., Geologic Evolution of Arizona, *Arizona Geological Society Digest,* Vol. 17.

Scalliet, B., F. Holtz, M. Pichavant, and M. Schmidt, (1996), "Viscosity of Himalayan leucogranites: Implications for mechanisms of granitic magma ascent, *Journal of Geophysical Research – Solid Earth, 101B,* pp. 27691-27699.

Schweitzer, M. and T. Staedter, (1997), "The real Jurassic Park," *Earth*, June, pp. 55-57.

Schwietzer, Mary H., Jennifer L. Whittmeyer, John R. Horner, Jan K. Toporski, (2005), "Soft-tissue vessels and cellular preservation in Tyrannosaurus rex," *Science*, Vol. 307, 25 March, pp. 1952-1955.

Science, (1984), Vol. 224, April 6, p. 3.

Science Daily, (2008), 25 September..

Selley, Richard C., (1970), *Ancient Sedimentary Environments*, Cornell University Press, Ithaca, NY, 237 pp.

Selverstone, J., (1982), "Fluid inclusions as petrogenetic indicators in granulite xenoliths, Pali-Aike volcanic field," *Contributions Mineralogy Petrology*, No. 79, pp. 28-36.

Shirley, S.B., (1991), "The Rb-Sr, Sm-Nd, andRe-Os isotopic systems," in Heaman, L. and J.N. Ludden, *Applications of Radiogenic Isotope Systems to problems in Geology,* Short Course Handbook, Vol. 19, 498 pp.

Shutong, X., A.I. Okay, and A.M.C. Sengor, (1992), "Diamond from the Dibic Shan metamorphic rocks and its implication for tectonic setting," *Science*, No. 256, pp. 80-82.

Skobelin, E.A., I.P. Sharapov, and A.F. Bugayov, (1990), "Deliberations of state and ways of perestroika in geology," *Critical aspects of the plate tectonics theory – Vol. I,* (Criticism on the plate tectonics theory), Theophrastus Publications, S.A., Athens, Greece, pp. 17-37.

Small, R.J., (1978), *The Study of Landforms: A Textbook of Geomorphology,* second edition, Cambridge University Press, London, p. 319.

Smith, D.C., (1984), "Coesite in clinopyroxene in the caledonites and its implications for geodynamics," *Nature,* No. 310, pp. 641-644.

Smith, P.V. Jr., (1952), "The occurrence of hydrocarbons in recent sediments from the Gulf of Mexico," *Science,* Vol. 116, 24 October, pp. 437-439.

Snelling, Andrew A., (1997), "Radioactive dating in conflict," *Creation ex Nihilo,* Vol. 20, No. 1, pp. 24-27.

Snelling, A.A. and Woodmorappe, J., (1998), "The cooling of thick igneous bodies on a young Earth; in: Walsh, R.E. (Ed.), *Proceedings of the Fourth International Conference on Creationism, Technical Symposium Sessions,* Creation Science Fellowship Inc., Pittsburgh, PA, pp. 527-545.

Snelling, Andrew A., (1999), "Iceland's recent mega-flood: An illustration of the power of Noah's Flood," *Creation,* Vol. 21, No. 3, pp. 46-48.

Snelling, Andrew A., (2000), "Conflicting ages of Tertiary basalt and contained fossilized wood, Crinum, Central Queensland, Australia", *Creation Ex Nihilo Technical Journal,* Vol. 14, No. 2.

Snelling, A.A., S.A. Austin, and W.A. Hoesch, (2003), "Radioisotopes in the diabase sill (upper Precambrian) at Bass Rapids, Grand Canyon, Arizona: an application and test of the isochron dating method," *Proceedings of the Fifth International Conference on Creationism,* Creation Science Fellowship, Inc., Pittsburgh, PA, USA, pp. 269-284.

Snelling, Andrew A., (2003), "The relevance of RB-Sr, Sm-Nd, Pb-Pb isotope systematics to elucidation of the genesis and history of recent andesite flows at Mt Ngauruhoe, New Zealand, and the implications for radioisotopic dating," in Proceedings of the Fifth International Conference on Creationism 2003, Creation Science Fellowship, Inc., Pittsburgh, PA, pp. 285-303.

Snelling, Andrew A. and Mark H. Armitage, (2003), "Radiohalos – A Tale of Three Granitic Plutons," in *Proceedings of the Fifth International Conference on Creationism 2003,* Creation Science Fellowship, Inc., Pittsburgh, PA, pp. 243-267.

Snelling, Andrew A., (2005), "Polonium Radiohalos: the model for their formation tested and verified," *Impact #386,* August, Institute for Creation Research; P.O. Box 2667; El Cajon, CA 92021, pp. i-iv.

Snelling, Andrew A., (2009), *Earth's Catastrophic Past: Geology, Creation & The Flood,* Institute for Creation Research; P.O. Box 59029; Dallas, TX 75229, Vol. 1, pp. 1-464, Vol. 2, pp. 465-1102.

Sonmor, Mark, (2005), "Dinosaur soft tissue: how long can it really last?," *Think & Believe,* Alpha Omega Institute, P.O. Box 4343, Grand Junction, CO 81502, Vol. 22, No. 3, May/June p. 3.

Southgate, P.N., 1980, "Cambrian stromatolitic phosphorites from the Georgia Basin, Australia, *Nature,* 285, 5 June, pp. 395-397.

Sparks, R.S.J., L. Wilson, and G. Hulme, (1978), "Theoretical modeling of the generation, movement, and emplacement of pyroclastic flows by column collapse," *Journal of Geophysical Research*, Vol. 83, No. B-4, pp. 1727-1739.

Steinen, R.P., (1978) "On the diagenesis of lime mud: scanning electron microscope observations of subsurface material from Barbadox, W.I.," *Journal of Sedimentary Petrology*, No. 48, p. 1140.

Stober, Dan, (2010), "The strange case of solar flares and radioactive elements," *Stanford Report,* 23 August, pp. 1-2. <ews.stanford.edu/news/2101/august/sun-082310.html>

Stokstad, Erik, (2005), "Tyrannosaurus rex soft tissue raises tantalizing prospects," *Science*, Vol. 307, 25 March, p. 1852.

Stokes, W.L., (1950), "Pediment concept applied to Shinarump and similar conglomerates," *Geological Society of America Bulletin*, Vol. 61, pp. 91-98.

Strabo, (1967), *Geography, Book XV: on India,* H.L. Jones (editor and translator), Harvard University Press, Cambridge, MA.

Sunderland, Luther D., (1988), *Darwin's Enigma: Fossils and Other Problems*, 4th edition, Master Books, Santee, CA.

Takahashi, K. et al., (1987), "Bound-state beta decay of highly ionized atoms," *Physical Review,* **C36**(4), pp. 1522-1527.

Taylor, B., A. Goodliffe, F. Martinez, and R. Hey, (1995), "Continental rifting and initial seafloor spreading in the Woodlark Basin," *Nature*, No. 374, pp. 534-537.

Taylor, Paul S., (1987), *The Great Dinosaur Mystery and the Bible*, Chariot Victor Publishing, Colorado Springs, CO, 63 pp.

Thiele, Edwin R., (1951), *The Mysterious Numbers of the Hebrew Kings,* Chicago, University of Chicago Press.

Thomas, Brian, (2013), "A review of original tissue fossils and their age implications," *Proceedings of the Seventh International Conference on Creationism,* Creation Science Fellowship, Pittsburgh, PA, USA, 20 pp.

Thomas, M.F., (1994), *Geomorphology in the Tropics: A Study of Weathering and Denudation in Low Latitudes*, John Wiley & Sons, New York, pp. 244-245.

Thompson, Alan Bruce, (1992), "Water in the Earth's upper mantle," *Nature*, Vol. 358, 23 July, pp. 295-302.

Thompson, Alan Bruce, (1999), "Some time-space relationships for crustal melting and granitic intrusion at various depths, in *Understanding Granites: Integrating New and Classical Techniques,* A. Castro, C. Fernandez and J.-L. Vigneresse eds., London: The Geological Society, Special Publication 168, pp. 7-25.

Tolan, T.L., S.P. Reidel, M.H. Beeson, J.L. Anderson, K.R. Fecht, and D.A. Swanson, (1989), "Revisions to the estimates of the areal extent and volume of the Columbia River Basalt Group," in: Reidel, S.P. and P.R. Hooper (editors), Volcanism and Tectonism in the Columbia River Flood-Basalt Province, *Geological Society of America Special Paper 239*, Boulder, CO, pp. 1-20.

Tosk, T., (1988), "Foraminifers in the fossil record: implications for an ecological zonation model," *Origins*, Vol. 15., pp. 8-18.

Toulkeridis, T., et al., (1998), "Sm-Nd, Rb-Sr, and Pb-Pb dating silicic carbonates from the early Archaen Barberton Greenstone Belt, South Africa," *Precambrian Research,* Vol. 92, pp. 129-144.

Twidale, C.R., (1981), "Origins and environments of Pediments," *Journal Geological Society of Australia*, Vol. 28, p. 425.

Tyler, David J., (1996), "A post-flood solution to the chalk problem," *Creation Ex Nihilo Technical Journal*, Vol. 10, No. 1.

Tyler, David J., (2006), "Recolonization and the Mabbul", in: Reed, J.K. and M.J. Oard (editors), *The Geologic Column – Perspectives within Diluvial Geology*, Creation Research Society; 6801 N. Highway 89; Chino Valley, AZ 86323, pp. 73-86.

Unknown, (1974), "The Earth's physical resources," *Block 3: Mineral Deposits*, Open University Press, Milton Keynes, England, p. 48.

Valley, J.W., J. McLelland, E.J. Essene, and W. Lamb, (1983), "Metamorphic fluids in the deep crust: evidence from the Adirondacks," *Nature*, No. 301, pp. 226-228.

VanDecar, J.C., D.E. James, and M. Assumpcao, (1995), "Seismic evidence for a fossil mantle plume beneath South America and implications for plate driving forces," *Nature*, Vol. 378, 2 November, pp. 25-31.

Vardiman, Larry, (1997), "Global Warming and the Flood", *ICR Impact #294*, December, pp. i-iv).

Velikovsky, Immanuel, (1955), *Earth in Upheaval*, Dell Publishing Co., Inc., New York, 288 pp.

Vonderheide, David, (2011-2013), personal communication.

Wakefield, R and G. Wilkerson, (1990), "Geologic setting of polonium radiohalos," in: Walsh, R.E. (Ed.), *The Second International Conference on Creationism*, Creation Science Fellowship, Pittsburg, Vol. 2, pp. 329-344.

Walker, Tas, (2001), "New radiohalo find challenges primordial granite claim," *TJ*, Vol. 15 (1), pp. 14-16.

Walker, Tas, (2004), "Peperite: more evidence of a large-scale watery catastrophe, *TJ*, Vol. 18, No. 1, pp. 18-21.

Walker, Tas, (2007), "Granite formation catastrophic in its suddenness," *Journal of Creation*, Vol. 21, No. 2, pp. 13-15.

Wang, X., J.G. Liou, and H.K. Mao, (1989), "Coesite-bearing eclogite from the Dabie Mountains, central China," *Geology*, No. 17, pp. 1085-1088.

Wang, Y., H. Xu, E. Merino, and H. Konishi, (2009), "Generation of banded iron formations by internal dynamics and leaching of oceanic crust," *Nature Geoscience*, 2, pp. 781-784.

Weisburd, Stefi, (1985), "The Earth's magnetic hiccup," *Science News*, Vol. 128, 5 October, pp. 218-220.

Wezel, Forese-Carlo, (1992), "Global change: shear-dominated geotectonics modulated by rhythmic Earth pulsations," in: Chatterjee, S. and N. Hotton III (editors), *New Concepts in Global Tectonics*, Texas Tech University, Lubbock, TX, pp. 421-439.

Whitcomb, John C. and Henry M. Morris, (1961), *The Genesis Flood,* Prespyterian and Reformed Publishing Company, Phillipsburg, New Jersey, 518 pp.

White, Errol, (1966), "A little on lungfishes," *Proceedings of the Linnaean Society of London*, Vol. 177, January, pp. 1-9.

Whitley, (1910), *Journal of the Philosophical Society of Great Britain, XII*, p. 56.

Whitmore, John, (2013), "The potential for and implications of widespread post-Flood erosion and mass wasting processes," *Proceedings of the Seventh International Conference on Creationism,* Pittsburgh, PA.

Wikipedia, (2012), "Neutrino," pp. 1-13).

Wile, Jay L., (2013), personal communication.

Wilson, Lionel and James W. Head, (1981), "Morphology and rheology of pyroclastic flows and their deposits, and guidelines for future observations," *The 1980 Eruptions of Mount St. Helens, Washington,* U.S. Geological Survey Professional Paper 1250, pp. 513-524.

Wise, Kurt P., (2003), "The Hydrothermal Biome: A Pre-Flood Environment," in *Proceedings of the Fifth International Conference on Creationism 2003,* Creation Science Fellowship, Inc., Pittsburgh, PA, pp. 359-370.

Wise, Kurt P., (2003), "The Pre-Flood Floating Forest: A Study in paleontological pattern recognition," in *Proceedings of the Fifth International Conference on Creationism 2003,* Creation Science Fellowship, Inc., Pittsburgh, PA, USA, pp. 371-381.

Woetzel, David, (2006), "The fiery Flying Serpent," *Creation Research Society Quarterly,"* Vol. 42, March, pp. 241-251.

Woodmorappe, John, (1978), "A Diluvian interpretation of ancient cyclic sedimentation" *Creation Research Society Quarterly*, Vol. 14, March, pp. 189-208.

Woodmorappe, John, (1980), "An anthology of matters significant to Creationism and Diluviology: Report 1," *Creation Research Society Quarterly*, Vol. 16, March, pp. 209-219.

Woodmorappe, John, (1982), "An anthology of matters significant to Creationism and Diluviology: Report 2," *Creation Research Society Quarterly*, Vol. 18, No. 4, March, pp. 201-223.

Woodmorappe, John, (1999), *The Mythology of Modern Dating Methods,* Institute for Creation Research; P.O. Box 2667; El Cajon, CA, 92021.

Woodmorappe, John, (no date), creation.com, "Billion-fold acceleration of radioactivity demonstrated in laboratory," pp. 1-3).

Woolley, A.R., (1989), "The spatial and temporal distribution of carbonatites," Bell, K. (ed.), Carbonatites: Genesis and Evolution, Unwin Hyman, London. pp.15-37.

Wright, G.F., (1911), *The Ice Age in North America and its bearing upon the Antiquity of Man*, (5th edition).

Young, Davis A., and Ralph F. Stearley, (2008), *The Bible, Rocks and Time*, IVP Academic, Downers Grove, Illinois, 510 pp.

Young, J.A., (2003), "Septuagintal verses Masoretic Chronology in Genesis 5 and 11", *The Fifth International Conference on Creationism - Technical Symposium Sessions,* Creation Science Fellowship, Inc.; Pittsburgh, PA, USA, pp. 417-430.

DEFINITIONS

Abiogenic – Not produced by living organisms.

Amorphous – Without definite crystalline structure.

Antediluvian Age – Time before the Worldwide Flood.

Basaltic magma – Alkaline, medium density, medium viscosity silicate magma that solidifies primarily into plagioclase. Forms a major component of Earth's modern oceanic crust.

Biogenic – Produced by living organisms.

Body fossil – A fossil exhibiting recognizable biogenic component parts, allowing identification as a specific complex lifeform.

Dilatacy – Expansion of deformed masses of granular material due to rearrangement of component grains.

Diluvial – Catastrophic flooding.

Equibrilibrium – In balance, unshifting, and unchanging.

Exsolutioned – The process of molten rock solutions separating from other cooling magma components.

Geomorphic – The origin and development of Earth's surface features.

Hydrodynamic – Rapid water flow eroding submerged rock.

Hypomagma – Acidic, low density, slow flowing, silicate magma that solidifies at lower temperatures than other magmas, primarily into orthoclase and quartz. Forms the majority of Earth's continental crust.

Isostatic – Vertical movement seeking to bring low and high density portions of Earth's crust into equilibrium.

Lineaments – Linear topographic features conforming to a vertical basement bedrock fracture network.

Metasomatic – Metamorphic alteration in which one mineral or mineral assembledge is replaced by another of different composition without melting.

Neocatastrophism – Most of the Rock Record was deposited intermittently during brief catastrophic events.

Orthogonal – Perpendicular linear seafloor topography features.

Permineralized – Fossilization by minerals deposited in pore spaces of originally hard animal parts.

Phanerozoic – The post-Precambrian portion of the Rock Record, characterized by body fossils and abundant evidences of life in sedimentary rocks.

Post-Flood Millennium – A time-frame of approximately one thousand years immediately following the Worldwide Flood.

Precipitate – Mineral solidification from liquid form.

Preoceanic – Prior to collapse of the ocean basin crust.

Proterozoic – (a.k.a. Precambrian) Geologic time between the Archaen and Paleozoic, which is characterized by the absence of body fossils.

Pyromagma – Alkaline, low density, fast flowing, low-silicate magma that primarily solidifies into halite, gypsum, calcite, and apatite. Contains abundant volatile liquid components, which either escape into the atmosphere or are encapsulated as vugs within the surrounding bedrock matrix. Proposed herein to have formed the vuggy Antediluvian crust at depths between 2 km and 25 km.

Quantized – Measurable increments of uniform spacing.

Recolonization – Establishing a unique new ecosystem.

Sapping – Seepage pressure in water saturated sediments that destabilizes the grain packing arrangement. Where groundwater surfaces down slope, sediment flow occurs and undermines overlying materials.

Sill - A tabular magmatic intrusion oriented parallel to parallel to planar bedding planes in the surrounding rock.

Subcrustal – Below and contacting the base of Earth's crust.

Tectonic – A movement or displacement of Earth's crust.

Tensional Fracturing – Pulling apart of Earth's crust.

Terrestrial – A subaerial or dry land habitat.

Tsunami – A laterally advancing wave generated by seafloor displacement.

Ultramafic magma – Alkaline, high density, slow flowing, silicate magma that solidifies primarily into pyroxene and olivine. Composes the majority of deep crust and upper mantle materials.

Unconformity – A stratigraphic surface of erosion.

Underplate – A regionally isolated magma body in contact with the base of Earth's crust.

Unfractured – An absence of cracks in bedrock, which acts as a barrier to fluid and gas migration.

Uniformitarianism – Observed slow and gradual depositional processes are applied to interpret unobserved depositional processes of the past.

Volatiles – Magma components remaining as liquids after other bedrock components have solidified.

Vugs – Tiny liquid filled cavities encased within a solid bedrock matrix.

www.ingramcontent.com/pod-product-compliance
Lightning Source LLC
Chambersburg PA
CBHW080240180526
45167CB00006B/2350